PHYSICS MADE SIMPLE

Christopher G. De Pree, Ph.D.
Agnes Scott College

A Made Simple Book
Broadway Books
New York

Produced by The Philip Lief Group, Inc.

PHYSICS MADE SIMPLE

A previous edition of this book was originally published in a different form in 1990 by Doubleday, a division of Random House, Inc., and is here reprinted by arrangement with Doubleday.

Printed in the United States of America

Produced by The Philip Lief Group, Inc.
Managing Editors: Judy Linden, Jill Korot.
Design: Annie Jeon.

Broadway Books titles may be purchased for business or promotional use or for special sales. For information, please write to: Special Markets Department, Random House, Inc., 1745 Broadway, New York, NY 10019.

MADE SIMPLE BOOKS and BROADWAY BOOKS
are trademarks of Broadway Books, a division of Random House, Inc.

Visit our Web site at www.broadwaybooks.com

First Broadway Books trade paperback edition published 2004.

Library of Congress Cataloging-in-Publication Data

De Pree, Christopher Gordon.

Physics made simple / by Christopher Gordon De Pree; produced by The Philip Lief Group, Inc.

p. cm. – (Made simple)

Rev. ed. of: Physics made simple / Ira M. Freeman. 1st ed. 1990

Includes index.

ISBN 0–7679–1701–4

1. Physics. I. Freeman, Ira Maximillian, 1905—Physics made simple. II. Philip Lief Group. III. Title. IV. Made simple (Broadway Books)

QC23.2.D4 2004

530—dc22 2004045855

10 9 8 7 6 5 4 3 2

To Phoebe,
who tried unsuccessfully to prevent me
from writing this book.

CONTENTS

INTRODUCTION

Physics made *simple*? Isn't that an oxymoron, like "work party" or "jumbo shrimp"? Not really. Although physics certainly can be a challenging discipline, requiring a deep understanding of difficult mathematics and years of experience and study, it is also a discipline that explains, in simple terms, how the physical world works—and that's where you come in.

Chances are if you have bought this book, you want to understand physics. Or perhaps you do not necessarily *want* to understand it, but you realize that you *need* to understand it as a means to an end. You may be enrolled in a noncalculus physics course and need some extra examples or a different approach. Or perhaps you had a bad experience in physics years ago (unfortunately a common experience, as I often hear on airplane trips when I say that I teach physics), and you want to return to the subject on your own terms. This book is written for the layperson who wants to acquire the basic tools required to explore and understand the physical world.

This book is meant for you either way, whether you are studying physics in an academic class, or studying it on your own. Only a basic knowledge of algebra and trigonometry is required, and the text is suitable as a companion volume for use with either a noncalculus- or calculus-based physics course.

Starting from some very basic principles of measurement, we will explore the more familiar topics of *motion, force, work and energy, sound, optics, electricity and magnetism,* and finally the less familiar world of modern physics: *quantum mechanics, relativity,* and *nuclear physics.*

Most of this book is spent describing physical models of the macroscopic world: balls, cars, trains, airplanes, planets, and so on. At the advent of the twentieth century, though, it became clear that the rules for very small objects (the size of the atom and smaller) and the rules for objects moving at very high speeds (close to the speed of light) must be slightly different from the rules that had worked for things about the size of humans moving at normal velocities. This understanding led to two of the great leaps in human understanding of the physical world: *quantum mechanics* and *relativity*. We will explore these fascinating topics in the last few chapters of the book.

The book is divided into six major parts: (I) Physics and the World, (II) Force, Work, and Energy, (III) States of Matter, (IV) Sound and Light, (V) Electricity and Magnetism, and (VI) Modern Physics. Each part is divided, generally, into three chapters.

Each chapter is laid out in a similar fashion. *Key terms* related to the major concepts in the chapter are listed at the opening of the chapter and are printed in italics within the text. Each chapter features two types of sidebars: one is titled **Explorations**, and the other is called **Physics in the Real World**. **Explorations** are simple experiments or calculations that you can try on your own to get experience doing what physicists do, observing and testing how the world really works. **Physics in the Real World** sidebars are explanations of everyday situations that physics can help you understand. For example, why does your plane have to reach a certain speed before it lifts off the runway (it has to do with *pressure*)? Or why do you need a tendon that connects to your heel (the Achilles tendon) for your foot to flex forward (it works like a *lever*)? At the end of each chapter there are **Problems.** These questions are designed to give you practice solving the kinds of problems that you might see in a physics course. The solutions to these questions are given at the back of the book.

I hope that reading this book will lead you to think about the world in the way a physicist does and will give you some insight into the reasons some people decide to make physics their life's work. Physics *can* be simple. You will see.

Let's begin the journey.

CHAPTER 1 PHYSICS: A FUNDAMENTAL SCIENCE

KEY TERMS

physics, chemistry, biology, biophysics, hypothesis, theory

UNDERSTANDING PHYSICS

Why study physics? For many students of physics, it is a means to an end and nothing else. A large number of premed students are required to show proficiency in physics to become medical doctors. Others need physics under their belt to become successful engineers. Yet others study physics on their way to being awarded degrees in biology, chemistry, architecture, environmental science, or astronomy; and there is the occasional student who, upon studying physics, decides that there is no other subject that will hold her interest like understanding fundamentally how the universe works.

So what is physics? In ancient times, the term *physics* referred to the study of the natural world and phenomena that took place in it. The study was sometimes called *natural philosophy*. A more modern definition of *physics* is that it is the study of matter and energy and the interactions between them, and the

PHYSICS IN THE REAL WORLD

As you sit in your mysteriously stained seat in an airplane waiting to take off, you hear the distinct sounds of the aircraft engines and systems turning on. The cabin door closes (creating a seal), and the pressure and temperature inside the aircraft can now be controlled from the cockpit. This control maintains a comfortable (and artificial) pressure and temperature inside the cabin even while you are cruising at 35,000 feet—an altitude at which the temperature and pressure outside the aircraft would be anything but comfortable.

Pressure and temperature both drop with altitude and are significantly lower at almost 7 miles (mi) up. The pilot says a few words about the travel time and aircraft speed on the way to your destination. Recalling that velocity times time gives distance, you make a quick calculation that home is about 500 mi away. The engines whine and the plane accelerates down the runway. At a certain speed, because of the profile of the aircraft wings, the pressure below the wing is greater than the pressure above the wing, and there is a net upward force exerted on the wing. The plane leaves the ground and you are on your way. The pilot guides the airplane to an altitude and speed such that the upward force on the wings—created by the pressure difference—is equal to the downward force due to the mass of your plane.

discipline of physics embodies a set of techniques to observe, model, and understand the natural world at its most basic level.

Over the millennia, scientists have found that objects seem to follow certain rules. A thrown ball follows a curved path through the atmosphere and back to the ground. An apple falling from a tree takes a certain amount of time to hit the earth. Careful observations of these *macroscopic*, or large-scale, phenomena have led physicists to formulate general rules about how things move and why they move in the way that they do. One of the distinctive aspects of any scientific discipline is that the models that scientists develop have predictive power; that is, any model that explains how things work should predict accurately how they will work the next time that the experiment is run, and should also be able to predict phenomena not yet observed. Thus, a physical model that explains where a thrown ball will fall should be able to predict where the ball will fall the next time it is thrown, perhaps at a slightly different velocity, or where a projectile other than a ball will fall.

It is important for the reader to keep in mind that physics, like any discipline, does not exist in a vacuum. It exists in relationship with all the other areas of human inquiry and understanding, and it is no better or worse than any of them. Certainly, a historian and a physicist look at the world in very different ways. If working at a small liberal arts college has taught me one thing it is that.

But even scientists in different disciplines look at the world in very different ways. A chemist might look at the Sun and be fascinated by the collections of atoms called *molecules* that are able to exist in the outer layers of the Sun's atmosphere. A physicist might look at the same Sun and marvel at the fact that the Sun's core is fusing 600 million metric tons of hydrogen into helium every second. An astronomer might reflect on the fact that the Sun has a diameter 109 times greater than that of Earth. In the case of a group of scientists, they all begin with the assumption that the best way to understand the physical world is to observe it, to model its behavior, and then to test the success or failure of these models.

PHYSICS IN RELATION TO THE OTHER SCIENCES

In a high school curriculum, physics is often taught last in a sequence that consists of biology, chemistry, and physics. In terms of the physical world, however, this order is somewhat backward. It would perhaps make more sense to study physics, then chemistry, and finally biology. *Physics* explains the interaction between matter and energy in the macroscopic world (objects the size of us, approximately) as well as the atomic and subatomic world. *Chemistry* builds on many of the theories of physics to explain the interactions among the atoms that form compounds and molecules, and it is fundamentally concerned with the properties of matter. There is significant overlap between the disciplines of physics and chemistry in our study of the structure of the atom. As one's studies of the atom move more into the structure and properties of matter, the details of interactions between compounds, and the formation of molecules and reaction rates, then one is leaving physics and moving squarely into the discipline of chemistry.

Certain chemical compounds (those containing carbon) are called *organic compounds*. The study of organic compounds is referred to as *organic chemistry*, and the gray area between chemistry and biology (sometimes referred to as *biochemistry*) involves the study of those particular molecules related to life processes. Once the scales of study are such that one is studying the smallest living things, then the discipline is generally considered to be *biology*. Biologists study all living things, and there are now overlaps between the disciplines of physics and biology that go beyond the molecular scale. In fact, one of the fastest growing disciplines in departments of physics is the field of *biophysics*, or the application of physics to biological phenomena. This hybrid field spans scales from the very small (the various microstructures that cells are able to make out of proteins) to the very large (the physics of flight in birds and insects, for example).

Another science related to physics is *engineering*, the general application of scientific principles to practical ends. The connection of physics to engineering is probably more obvious. Engineers build structures, among other activities, and these structures need to be able to withstand their own internal forces as well as those occasional forces that they may be subject to (e.g., earthquakes or wind shear). A thorough understanding of forces is essential to engineers, who then add specialized understanding of the properties of materials to design objects and structures that can survive in the physical world. For these reasons, and many others, a fundamental understanding of physics is useful to scientists in many disciplines, and a basic physics course is generally required of future chemists, biologists, and engineers.

PHYSICS IN RELATION TO NONSCIENTIFIC DISCIPLINES

There are many aspects of physics that tie it to any number of nonscientific disciplines. The physics of sound, of course, is critical to the understanding of musical instruments and the manipulation of sound by musicians. Artists, for example, often use the physics of light to work with colored glass to create light sculptures. The fundamental nature of the questions posed by physicists about the way the world works places them in contact, from time to time, with theologians asking many of the same questions: How did we get here? Are there universal laws, and if so, how did they arise? Does the presence of universal laws imply the presence of a creator, or are the laws just a fundamental part of this universe? In addition, many mathematical techniques that are useful to physicists interested in modeling complex physical systems are just as useful to economists modeling complex systems in the marketplace.

PHYSICS AND HUMAN CURIOSITY

Our understanding of the world derives from the sciences and the humanities equally. Physics is a part of a continuum of human curiosity; but as you will learn in this book, physicists take a unique view of the world. Physicists observe the physical world to formulate *hypotheses*. Further observations either confirm or refute hypotheses; physicists have to keep open minds and be willing to revise, or, if necessary, throw out old ideas and formulate new ones.

Any hypothesis worth its weight can and will be tested by further observations, perhaps by other investigators. A hypothesis that has survived multiple tests and can be used to make predictions about other natural phenomena is referred to as a *theory*.

In this book, we begin our exploration of physics with the most familiar topics and progress to topics that may seem more and more foreign. We start with an explanation of measurement and motion, and end with an overview of the fundamental particles that make up our universe, things with names like *quarks* and *gluons*. It is important to remember that all this knowledge is connected.

The same curiosity that made early societies begin to take, quite literally, a measure of their world has led humans to explore the very stuff that makes up the universe. The same consciousness that first noticed the regularity of the motions of the moon and developed mathematics has more recently designed and constructed enormous subterranean tanks to detect nearly massless particles called *neutrinos* produced in fusion reactions in the core of the Sun. The same instinct for order that first defined the meanings of inch and pound or meter and kilogram has postulated the presence of *dark matter*, a material with mass but unknown composition that apparently fills the universe.

We never know exactly where our observations and our hypotheses will lead us, and that is part of the excitement of physics.

CHAPTER 2

MEASUREMENT AND ESTIMATION

KEY TERMS

fundamental units, mks system, derived units, weight, mass, area, volume, density, scientific notation, significant figures, order of magnitude calculation

PHYSICS IN THE REAL WORLD

You may wonder about the relevance of units of measurement to your life. What could be the possible use, you might ask, of converting quantities from feet to meters and vice versa? Scientists at NASA and Jet Propulsion Laboratories in Pasadena, California, recently discovered just how relevant units can be. A failure to convert engine thrust values from English to metric units caused a NASA mission to Mars to fail catastrophically, plowing the probe into the planet's surface. Here is a quote from the report explaining the loss of the Mars Climate Orbiter (MCO) spacecraft: "the root cause for the loss of the MCO spacecraft was the failure to use metric units in the coding of a ground software file, 'Small Forces,' used in trajectory models." A trajectory is a path followed by any projectile, be it a baseball or a spacecraft. The failure to use correct units by these spacecraft teams resulted in the loss of the $328 million mission. So read this chapter carefully.

UNITS OF MEASURE

Physics, like much of science, is all about measurement. Since ancient times humans have figured out ways to take a measure of their world. How many sheep do I own? How large is my tract of land? How many days until the winter festival? How much does this steer weigh? Of course, measurement is fundamental to any science, but particularly to physics, the science that probes the physical properties of the natural world from the scale of the atom to the scale of clusters of galaxies.

We begin here with an explanation of the basics of measuring length, mass, and time. Physicists use the *mks system*, which stands for meter-kilogram-second, along with the *fundamental units* of length, mass, and time. We will use these fundamental units in various combinations to explore many other aspects of the physical universe. (Astronomers, by the way, use the cgs system, short for centimeter-gram-second.) *Derived units* are simply combinations of fundamental units. An example of a derived unit is meters per second (m/s), which is used to measure velocity.

Some quantities are more easily measured than others. For example, the size of a room can be measured with a tape measure, and you can measure your weight with a scale. The diameter of a galaxy, however, must be measured in a different way, involving knowledge about the distance to the galaxy and its apparent size in the sky. The mass of an *electron* cannot be measured with a scale, but (in one method) the mass is calculated from the force exerted on electrons moving in a magnetic field. Various equations give the acceleration, velocity, and kinetic energy of the electrons as they are stripped from a heated filament in a tube and pass through a hole in the anode of the tube. The mass is indirectly determined from measurements of the current, voltage difference, and magnetic field. The sizes of atoms and molecules have been measured not with rulers but with the use of X-rays. The masses of the Sun and planets in the solar system have been measured using Kepler's laws of planetary motion.

When making measurements of the physical world, there are important choices to be made. The size of the unit chosen should be appropriate to the object being measured. Therefore, astronomers do not typically use the centimeter to discuss the distances to other stars; they instead use the *light-year*, or the distance a photon moving at the speed of light (3×10^8 m/s) can travel in a year. (A *photon* is a "packet" of radiation. Light or other transmitted electromagnetic radiation, such as radio waves, microwaves, or x-rays, are all transmitted as photons.) Likewise, nuclear physicists do not use acres or hectares to measure the apparent size of atomic nuclei; they use barns. A *barn* (a unit chosen with some sense of humor) is defined as 10^{-28} m^2, a tiny area.

PHYSICS NOTE: By the way, a new unit of measure can potentially earn a physicist everlasting fame. Many units of measure used in physics and in everyday life are named in honor of the scientists who first devised them, including the hertz (Hz), a measure of electromagnetic frequency (named after Heinrich Hertz, 1857–1894), and the joule (J), a unit of energy (named after James Prescott Joule, 1818–1889).

Making useful measurements in physics and astronomy is often about using the right unit to make your job easier; however, when astronomers and physicists calculate derived quantities, like the energy production rate of a star, they always revert to their fundamental units (cgs or mks).

In 1960, an international committee of scientists devised a standardized system of designations for these fundamental quantities. This system is known as the Système International, or SI.

Table 2.1 lists the Système International (SI) fundamental units and their abbreviations.

Quantity	Unit	Abbreviation
Length	meter	m
Mass	kilogram	kg
Time	second	s

Table 2.1 Fundamental Units of the Système International (SI)

THE MKS SYSTEM

The measurement of length, or distance, is something that we encounter daily. The odometer on your car ticks off the miles or kilometers. The doctor measures your height in feet and inches at your annual checkup. You count off the 25-m laps that you are swimming at your local pool.

As mentioned previously, the unit used for measurement of a particular length or distance should be based on the size of the area being measured. Therefore, the English system has small units (inches) and large units (miles) to represent length; however, the relationship between various units of length is somewhat arbitrary: 12 inches (in.) in a foot (ft), 3 feet in a yard (yd), and 1760 yards in a mile (mi). One of the original motivations for the metric system (adopted first by France in the late eighteenth century) was to get rid of these arbitrary ratios and to replace them with powers of 10. Thus, there are 10 millimeters (mm) in a centimeter (cm), 100 cm in a meter (m), and 1000 m in a kilometer (km). The metric system is standard for all scientific work, and in the engineering and manufacturing industries outside of the United States.

PHYSICS NOTE: The United States is almost alone in the world in persisting in its use of the English system of measure (inches, feet, gallons, etc.). In the English system, the standard yard was taken to be the distance between the end marks on a bronze bar kept in a vault at the Office of the Exchequer in London. In the modern era, however, the yard is defined in terms of the meter (1 yd = 0.9144 m). Although we will occasionally refer to the English system of units in this book for familiarity, we will generally use the metric system for measurements of length and mass. Happily, the English and the metric systems use the same fundamental unit for time (the second).

Length

The fundamental unit of length in the metric system is the meter (m). Originally defined as one ten-millionth of the distance from the equator to the North Pole in 1799, the standard meter was internationally defined as the length of a bar of platinum alloy kept at the International Bureau of Weights and Measures in France. Bureaus of Standards in other countries kept precise copies of this bar. The meter is slightly longer than the yard (39.37 in.). In 1983, the meter was redefined to be the distance traveled by light in a vacuum in 1/299,792,458 of a second.

The utility and simplicity of the metric system can again be seen in the conversions from one multiple of length or mass to another (see Table 2.2). For example, there are 1000 grams (g) in a kilogram (kg). There are 100 cm in a meter. A nanometer is one-billionth of a meter, and so on. Table 2.2 indicates some of the most common metric prefixes used in the metric system. Spaces in the table are inserted where you might be used to seeing commas; these separations just aid in the readability of the numbers. Multiples of any fundamental units are simply named by placing a prefix before the word *meter*. A *kilo*meter, for example, from the Greek word for "thousand," is 1000 m.

Multiple	Prefix
1 000 000	mega (M)
1 000	kilo (k)
0.01	centi (c)
0.001	milli (m)
0.000 001	micro (μ)
0.000 000 001	nano (n)

Table 2.2 Common Metric Prefixes

Mass

Students often confuse the terms *weight* and *mass*. The *weight* of an object depends on two things: its mass and the gravitational force pulling on it. You may remember hearing as a child that your weight would be less on the Moon. That is true, because the Moon exerts a smaller pull on your mass than does Earth. The mass of an object, however, does not vary. *Mass* is fundamental, because it is a measure of the number of atoms in an object. Your mass on the Moon, on Earth, or wherever you want to picture yourself in the universe remains unchanged.

The fundamental unit of mass is the kilogram (kg). As with metric measures of length, metric measures of mass differ from each other by powers of 10. There are 1000 milligrams (mg) in a gram (g) and 1000 g in a kilogram (kg), for example.

Although *weight* cannot be used as a fundamental measurement (since it varies with your location in the solar system), it is a concept that we are comfortable with. Most people know their weight to within a few pounds, whereas they have no idea what their mass is. We can easily convert between the two if we know where the weight is being measured. For example, at the surface of Earth, 2.21 pounds (lb) is the force exerted on an object with a mass of 1 kg. Thus, to calculate your mass in pounds, just divide your weight (in pounds) by 2.21. The result will be your mass in kilograms. My mass is 83.7 kg. Determining my weight is left to the interested reader.

With humans, planets, and stars, of course, mass can change. If you eat a lot of cream pies, you will add to the number of atoms that your body comprises (stored as fat), and your mass (and weight) will increase. Stars lose mass slowly over time through a phenomenon known as *stellar winds*.

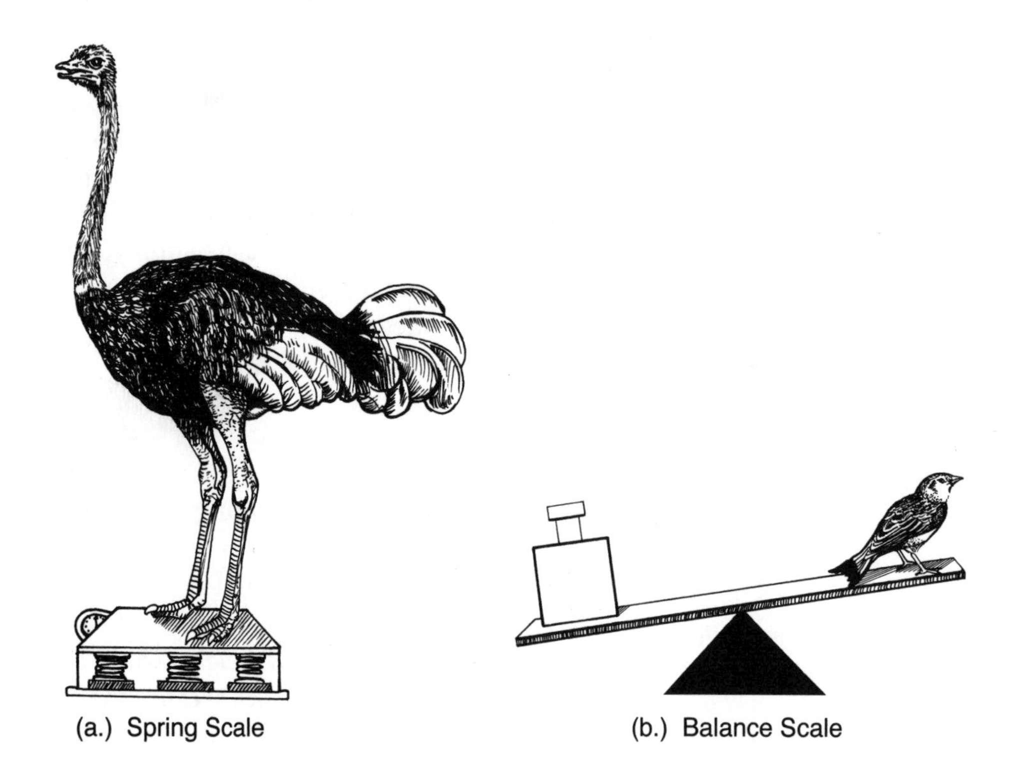

Figure 2.1
Spring scales that we use day-to-day tell us our weight (measured in pounds). Balance scales can tell us our mass (in kilograms) by comparing our mass with known masses.

Time

The passage of time is certainly something fundamental and a quantity that we can measure, but time itself is a quantity that is rather difficult to define. Ancient societies noticed that motions of objects in the sky (the Sun, the Moon, the stars, and the planets) could be related to the passage of time. The first definition of the fundamental unit of time, the second, was 1/86,400 of a solar day. This is because there are $60 \times 60 \times 24 = 86{,}400$ seconds in a day. A *solar day* is the elapsed time between two successive solar crossings of the meridian, or the line that divides the sky in half (east and west). When the Sun is *ante meridian* we are in the A.M. part of the day, and after the Sun crosses the meridian, we are *post meridian* (P.M.).

In both the English and the metric systems, the basic time unit is the *second* (s). Multiples of time have never been transformed into the metric system (we do not have 100 s in an hour, for example, but rather 60), so units of time have retained a certain oddness: 24 hours (h) in a day, 365.25 days in a year. These last two examples, of course, are set by two random periods of time, how long it takes Earth to rotate once on its axis and the time it takes for Earth to orbit the Sun. In the modern day, the *second* is defined in terms of the number of oscillations of an isotope of the element cesium (cesium-133): a second comprises 9,192,631,770 oscillations. Scientists do use metric multiples of fractions of a second, however, with times measured in milliseconds, nanoseconds, and so forth. For example, there are stars called *millisecond pulsars* that rotate 1000 times every second.

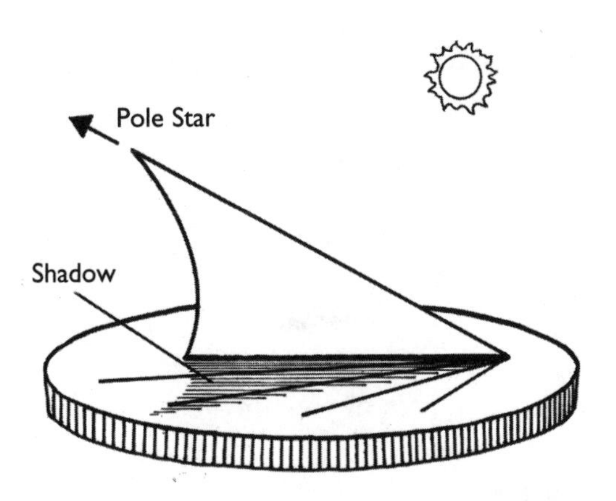

Figure 2.2
This schematic shows the basic parts of a sundial. The sundial has been used for thousands of years to tell time.

AREA, VOLUME, AND DENSITY

Area

One of the simplest ways to think about area is to think of tiles. If you have a standard tile (let's say, 1 cm on a side), how many tiles would it take to cover the region you are interested in measuring? The answer is the *area* of the region in "square centimeters." A square tile 1 cm on a side has an area of one square centimeter (abbreviated 1 cm^2). Area is a two-dimensional quantity that can be determined (for the simplest shapes) by taking the product of the length and width of a defined rectangular region:

A (area) = L (length) $\times$ W (width)

Thus, a region that is 2 cm by 4 cm has an area of 8 cm^2. Another way to think about this measurement is that it would take 8 of our 1-cm^2 tiles to cover such an area. Obviously, there are irregularly shaped regions that would not be covered by an exact number of tiles. Such regions, like a small puddle in the road, might have an area of 10.23 cm^2. The areas of other regular shapes can also be determined quickly by using formulas for area. To deter-

EXPLORATION 2.1

You are about to replace the sod in your front yard, a rectangular area that is 60 ft wide and 20 ft deep. If the sod comes in squares that are 18 in. on a side, how many squares will you need to do the job? If the sod comes in pallets of 50, how many pallets will you need to order?

Answer:
The area that you need to sod is easily computed by multiplying the width times the height: 60 ft $\times$ 20 ft = 1200 ft^2. If the sod is 18 in. on a side, or 1.5 ft on a side, then each square has an area of 1.5 ft $\times$ 1.5 ft, or 2.25 ft^2. To find the number of squares you need, you divide the total area by the area of each square, or 1200 ft^2/2.25 ft^2, giving us $533\frac{1}{3}$ squares of sod. You will need 534 squares, or 11 pallets (you will actually need fewer than 11 pallets, but we assume that you cannot buy partial ones).

mine the area of a circle, for example, you must first know the distance from the center of the circle to its edge, or half its diameter. This is called its *radius* or (r). So the area of a circle can be calculated as:

$$A \text{ (area)} = \pi r^2$$

where π (pi) is a shorthand way to refer to the ratio of a circle's circumference to its diameter. The value of pi is approximately 3.1415, but it can be calculated out to an infinite number of decimal places.

Volume

Volume is a measurement of how much space something occupies in three dimensions. We can consider volume much like we just thought about area. This time, however, instead of a tile, imagine a tiny cube that is 1 cm on each side, like a die. The volume of an object can thus be thought of as the number of these tiny 1-cm-sided dice it would take to occupy the same amount of space.

We can easily determine the volume of objects with perpendicular sides as follows:

$$V \text{ (volume)} = L \text{ (length)} \times W \text{ (width)} \times H \text{ (height)}$$

Therefore, each of our 1-cm-sided dice has a volume of 1 cm $\times$ 1 cm $\times$ 1 cm = 1 cm^3. Thus area (a two-dimensional quantity) is measured in square centimeters (cm^2), and volume (a three-dimensional quantity) is measured in cubic centimeters (cm^3).

EXPLORATION 2.2

You can measure the volume of an irregularly shaped object, such as a rock, by immersing it in water in a graduated container and measuring the rise in the level of the fluid. The volume of the object will be the same as the volume of the displaced water.

Find a rock outside in your yard. (Don't use a porous rock like lava or sandstone, as the results will be different.) To measure the volume of the rock, fill a glass measuring cup with water, place the rock in the cup, and determine the change in the level of the water. You will use this value (in milliliters) for your volume. Use a kitchen scale to determine the mass of your rock in grams. Calculate the density of your rock in grams per cubic centimeter (1 cm^3 = 1 mL). Does your rock have a density similar to any of the values in Table 2.3?

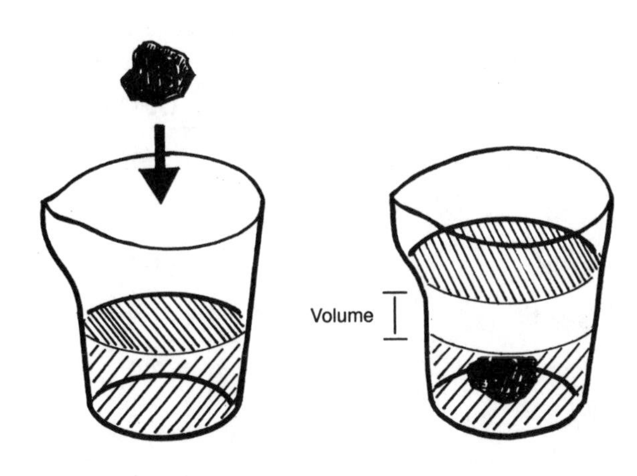

Figure 2.3
Measuring the volume of an oddly shaped object (like a rock). The change in the apparent volume of the water is due to the rock.

The volume of 1000 cubic centimeters (cm^3) has been defined to be a liter (L). To have a volume of 1000 cubic centimeters, an object could be 10 cm on a side, so that 10 cm × 10 cm × 10 cm = 1000 cm^3. Ten cm is the same as 0.1 m; therefore, a liter is 0.1 m × 0.1 m × 0.1 m, or 0.001 m^3.

We often buy drinks in 2- and 3-L-sized bottles. A liter is just slightly larger than the English quart. A subdivision of the liter, the milliliter (mL), is a unit often used for the administration of oral medicines. A popular drink in the United States, milk, is still typically purchased in the English volume units: gallons and half-gallons. Another popular liquid in the United States, gasoline for cars, is purchased by the gallon. The rest of the people in the world (even the English) buy their gasoline by the liter.

Other more complicated shapes have well-defined volumes. For example, the volume of a sphere of radius r is

$$\text{Volume} = \frac{4}{3}\pi r^3$$

and the volume of a circular cylinder (think of a can) of radius r and height h is

$$\text{Volume} = \pi r^2 h$$

Appendix E gives formulas for calculating the area and volume of a number of regular shapes.

Density

Density is a measurement of how much mass is packed into a given volume. In B-movies of the 1950s, Hercules would toss large rocks through the air, fending off attackers or simply impressing the ladies. Of course, the actor playing Hercules couldn't really toss large rocks around. The fake rocks were large in volume and painted to look like rocks; but in fact, they were made of a very low-density material: Styrofoam. Thus, although they had the same volume, shape, and color as rocks, they had the mass of, perhaps, a baseball; so the actor could toss the "rock" through the air and appear to have the strength of many men.

The *density* of an object is defined to be its mass divided by its volume, and so in the metric system, the density of an object is measured in kilograms per cubic meter, or in grams per cubic centimeter. Solids, liquids, and gases all have different densities. The density of oil is greater than that of water, for example, which causes oil and water to separate into two layers, with the less dense water floating on top. The planets in the solar system have different densities, too, with the terrestrial planets (like Earth) being more dense than the Jovian planets (like Jupiter). In general, density is calculated as:

$$D \text{ (density)} = M \text{ (mass)}/V \text{ (volume)}$$

The density of water is 1000 g per 1000 cubic centimeters (cm^3) or, more simply, 1000/1000 = 1 gram per cubic centimeter (g/cm^3) at a temperature of 4°C. Table 2.3 shows the densities of some common materials in cgs units of grams per cubic centimeter.

EXPLORATION 2.3

A king is afraid that his pure gold crown has been stolen and replaced with a gold-lead alloy crown. Using the density of gold (see Table 2.3), determine the volume of a 2-kg pure gold crown. If the crown is made of a gold-lead alloy, will the volume be greater or less than this value. Why?

Answer:
Density is equal to mass divided by volume. The mass of the crown is 2.0 kg. The density of gold is 19.3 g/cm^3, or 0.0193 kg/cm^3. The volume of a pure gold crown would then be 2 kg divided by 0.019 kg/cm^3, or 1.0×10^2 cm^3 (at two significant figures). If the crown were made of pure lead, its density would be 0.0113 kg/cm^3, and the volume would be 180 cm^3. Thus, a gold-lead alloy crown would have a greater volume than a pure gold crown.

By simply rearranging the density formula, you may, of course, solve this equation for the mass or volume of an object. Thus, if you know any two of the properties of a substance, you can find the third:

$$M = D \times V$$

or

$$V = M/D$$

Substance	Density (g/cm^3)
aluminum	2.7
iron	7.90
lead	11.3
gold	19.3
limestone	3.2
ice	0.92
wood, pine	0.5
gasoline	0.70
water	1.00
seawater	1.03
mercury	13.6
air	0.0013
hydrogen	0.00009

Table 2.3 Densities of Some Common Materials

SCIENTIFIC NOTATION

Because of the enormous range in scales that physicists deal with, they (and you) need to become familiar not only with measurement but with the use of a shorthand known as *scientific notation,* a way to represent very large and very small numbers in a compact way. This shorthand is also sometimes called *powers of 10* notation, because it entails thinking of large and small numbers as multiples of the number 10. Referring to multiples of 10 for various measurements fits perfectly with units of measurement in the metric system. For example, there are 1000 (or three powers of 10, 10^3) millimeters in a meter.

A number written in scientific notation has two parts: (1) a coefficient, which is a number between 1 and 10, and (2) the power of 10, which is an exponent, either positive or negative. A positive exponent means that the number is greater than 1, and a negative exponent means that the number is less than 1. An exponent of zero gives 1 ($10^0 = 1$). Thus, the format of a number in scientific notation is

$A \times 10^B$

Table 2.4 gives some examples of numbers written both as decimals and in scientific notation.

As we have said, the fundamental unit of length is the meter. One million meters is 1,000,000 m, or six powers of 10, written as 10^6 m; that is, 10^6 m is a 1 followed by 6 zeros. The exponent indicates how many places to the right of the coefficient value to move the decimal. For example, 5.38×10^2 m is the same as 538 m, because we moved the
decimal two places (two powers of 10) to the right.

The astronomical unit (AU) is the mean distance between Earth and the Sun, or about 1.5×10^{11} m. This number is much easier to write and manipulate in scientific notation than the number 150,000,000,000 m. A light-year (LY), or the distance that light travels in a year, is 9.5×10^{15} m. How many times would you want to write out 9,500,000,000,000,000 m?

Number	Scientific Notation
10,000 m	1×10^4 m
538 s	5.38×10^2 s
0.000 000 023 kg	2.3×10^{-8} kg
0.05 kg/m^3	5×10^{-2} kg/m^3

Table 2.4 Examples of Scientific Notation

The same shorthand can be used for numbers that are very small. A millimeter is 0.001 m, or 10^{-3} m. When the number is smaller than 1, the exponent indicates how many places to the left of the coefficient to move the decimal. A micron is a millionth of a meter. You could write this as 0.000001 m or, more simply, as 10^{-6} m.

Multiplication and Division

Multiplying and dividing numbers in scientific notation is also relatively simple. If you are multiplying two numbers in scientific notation, you multiply the coefficients and add the exponents. For example, if you know that a sample of mercury has a density of 13.6 g/cm^3 and a volume of 200 cm^3, then you could determine the mass to be (using $V \times D = M$):

$2.0 \times 10^2 \text{ cm}^3 \times 1.36 \times 10^1 \text{ g/cm}^3 = 2.7 \times 10^3 \text{ g}$

Dividing two numbers in scientific notation involves dividing the coefficients and subtracting the exponents. If you want to add or subtract two numbers in scientific notation, it is important to convert them to the same units before doing so. For example, if we wanted to add 1 cm to 1 m, we could *not* say that 1 cm + 1 m = 2. What units would we use? We would need to add 1 m to 0.01 m, giving us 1.01 m, or 101 cm. In scientific notation, this would be 1.0×10^2 cm + 1.0 cm, giving 1.01×10^2 cm.

Significant Figures

These examples bring us to the topic of significant figures. *Significant figures* are those digits in a final calculation that have physical meaning. If we are measuring quantities, the number of significant figures is determined by the accuracy of our measurements. For example, a ruler with centimeter divisions would not give us as accurate a measure of length as a ruler with millimeter divisions. For performing calculations the rules are as follows:

1. For multiplication or division, keep only as many significant figures as there are in the least accurate of the numbers being multiplied or divided. Thus, in the preceding example, when we multiply 2.0×10^2 cm^3 by 1.36×10^1 g/cm^3, we retain only two significant figures, giving us an answer of 2.7×10^3 g.

2. For addition and subtraction, we retain the smallest number of decimal places that appears in the numbers being added or subtracted. Therefore, if we add 1.00 m and 2.1 m, our answer is correctly 3.1 m, *not* 3.10 m.

ORDER OF MAGNITUDE CALCULATIONS

Physicists often make quick calculations that are approximately correct. These are called *order of magnitude calculations*. Enrico Fermi (1901–1954) was probably the most famous physicist fond of making such calculations. In these types of calculations, we round all the appropriate input values to the nearest power of 10 and then multiply or divide the input values to arrive at an answer that is informative, if not exact. These types of calculations often help physicists evaluate whether they are approximately correct in their understanding of a problem before they work out the exact answer.

As an example, what if we wanted to determine the total amount that Americans spend on soda each year? Well, there are about 300 million Americans, or 3×10^8 people. An average American, let's say, consumes two sodas each day, at a cost of $1 each. There are 365 days in a year (let's call it 400 days). We then multiply all these factors:

(3×10^8 people) (4×10^2 days/year) (2 sodas/day/person) ($1/soda) = 24×10^{10} per year, or $240 billion per year.

Given that a quick check on the Internet indicates that Americans spend over $30 billion in vending machines alone, this value is probably a reasonable estimate; and several hundred billion dollars is a lot of money.

What if we wondered about the distance covered if every man, woman, and child in

EXPLORATION 2.4

Using an order of magnitude calculation, determine the number of hours of television watched (in total) by all Americans in a year. Convert this number to days and years.

Answer:
Let's assume that Americans watch approximately 3 h of television per day. Multiplying this by the approximate number of days in a year (400), we get a total of 1200 h. Dividing this by 24 h in a day gives the equivalent of about 50 days of television. Assuming that there are approximately 300 million Americans in the country, this means that Americans watch (50 days/American) × 300 million Americans, or 15 billion days of television each year, or over 40 million years of television!

the United States were to lay head to toe. Again, there are about 3×10^8 Americans, and (let's say) their average height is 1 m. Therefore, their collective height is

$$3 \times 10^8 \text{ people} \times 1 \text{ m/person} = 3 \times 10^8 \text{ m}$$

This is almost exactly the distance that light travels in 1 s. The speed of light is 3×10^8 m/s. This distance is also the approximate mean distance between Earth and the Moon. If every man, woman, and child in the United States was to stand on one another's shoulders, the resulting people ladder would reach nearly to the Moon (barring other difficulties).

PROBLEMS

2.1 Convert your height in feet and inches to a height in meters and centimeters.

2.2 Determine the surface area of Earth (assuming it's a sphere) from its radius, which is about 6400 km. What is the area of Earth's oceans if they cover 2/3 of Earth's surface?

2.3 Write the following numbers in scientific notation: (a) 299,792,458 m/s, (b) 86,400 s, (c) 9,192,631,770 oscillations, (d) 0.000 015 cm.

2.4 What is the volume of a sphere with a radius of 10 cm? If the sphere has a mass of 10 kg, what is its density? If the sphere is solid, what would you guess its composition to be?

2.5 How many millimeters are in a kilometer? How many centimeters are in a millimeter?

2.6 How many seconds are in a day? In a month? In a year?

NOTE: The last three problems, like the examples worked in the chapter, have answers that will depend on the assumptions you make in solving them. Your estimates may be slightly different, but should have the correct order of magnitude (power of 10).

2.7 Make an order of magnitude calculation of the number of gallons of gas used by Americans to run their cars every year.

2.8 Make an order of magnitude calculation of the amount of money Americans spend on fast food each year.

2.9 Make an order of magnitude calculation of the number of cars on the streets of the United States.

CHAPTER 3

OBJECTS IN MOTION

KEY TERMS

position, velocity, acceleration, speed, vector, scalar, resultant velocity, resultant, free fall

Everything around us is in constant motion. Airplanes fly overhead, cars whisk down the street, children throw a ball back and forth in a park, even molecules of oxygen, nitrogen, and carbon dioxide move about in the atmosphere in a constant jittery dance. All these phenomena involve motion. In this chapter we explore the ways in which physics describes the motion of objects through space. Although most of the examples in the chapter involve large (macroscopic) objects, many of the same rules apply to very tiny objects as well. Using two fundamental quantities, position and time, we can describe where an object is, where it is headed, and how long it will take to get there.

POSITION, VELOCITY, AND ACCELERATION

There is interplay between the position, velocity, and acceleration of an object. Its *position* is simply an object's location in space (often indicated by the letter x). Its *velocity* (v) is the rate of change in its position, and its *acceleration* (a) is the rate of change of its velocity. We will consider each of these important measured quantities in turn.

Position

One fundamental aspect of an object that we can measure is its position. Position is an inherently "relative" measurement. We might say, "I am two miles south of school," or "I live in an apartment that is three blocks west of the park." Measurements of position require a zero point, and the zero point that we choose is, of course, arbitrary. In a laboratory experiment, you might be measuring the motion of a cart from a resting state. The position of a cart at the top of an inclined plane at the start of an experiment is its zero point; its motion (down the incline) is measured from that arbitrary zero point.

Many of us are most familiar with position in one dimension: a number on a number line has a position, or distance from zero. We usually use the variable x to indicate position in one dimension. In two-dimensional space, there are many choices for variables indicating position, but we generally use the Cartesian plane (named after French philosopher and mathematician Renèe Descartes [1596–1650]), or the x-y plane. The x-y plane is used frequently in problems in this book, because we can pinpoint an object in two-dimensional space with no more than two coordinates. In three-dimensional space, three coordinates are required, and we add one more letter to the list and specify the x-, y-, *and* z-coordinates of an object.

Figure 3.1
In an aerial photograph of Nebraska, the "grid" of fields and roads is apparent. Many Western states have roads laid out in grids that are aligned with the cardinal points of the map. This grid is an example of a Cartesian grid. (Image courtesy of USGS)

In studies of motion, we are concerned with the measurement of the change in position of an object. For example, we might want to know that in a specified period of time, an object covered a measured distance. The position of an object, and how it changes as a function of time, are basic measurements that we can make, and we can use these measurements to test our theories of how objects move through space. Distances are typically measured in meters or multiples of this unit (e.g., centimeters, kilometers).

One early observation, made by Italian Scientist Galileo Galilei (1564–1642) and others, was that an object dropped from a great height fell to Earth in a very regular way. That is, an object dropped from a certain height above the ground seemed to take a constant amount of time to strike Earth, and that (in careful experiments) the time required did not depend on the mass of the object.

In the centuries before Galileo changed the world with his experiments, the ideas of the philosopher Aristotle (384–322 B.C.E.) carried a lot of weight. Aristotle thought about how the world should work, often without the burden of experimentation or checking how they actually were. Aristotle, for example, proposed that pushes or pulls were required to keep objects in motion.

Aristotle also proposed that heavier objects fell faster than lighter objects. Galileo, on the other hand, did not simply propose what he thought *should* happen; he went out into the world and observed it. After careful observations, Galileo found that heavier objects *in fact* do not fall any faster than lighter ones. It is just that air resistance has a more obvious effect on light objects (like feathers) than heavy ones (like hammers). Galileo was also one of the first scientists to explore the nature of friction and the effect that it has on the motion of objects. He concluded that in the absence of air friction, objects with different masses would fall to Earth at the same rate. Galileo also concluded that objects do not, in fact, require pushes or pulls to keep them going; they will go on forever without any additional help if left alone. In the real world, of course, friction cannot be avoided.

One of Galileo's great contributions to physics was the exploration of objects in motion, which was made possible by the accurate measurement of time. Once time could be measured accurately, then not only the position but also the velocity and acceleration of an object could be measured. The regularities measured in the acceleration of objects at the surface of Earth eventually led to the discovery of gravity, a topic we will cover in Chapter 6.

Speed and Velocity

People often use the terms *speed* and *velocity* interchangeably, but in physics they are two distinct quantities. Earlier in the chapter, *velocity* was defined as the rate of change in position. In properly measuring the motion of an object, however, we must specify two quantities: how fast it is moving (the rate of change in its position) and in what direction. How fast an object moves (measured in meters per second or in miles per hour) is its *speed*.

Speed = Distance/Time

or

$s = d/t$

You probably use this formula all the time. For example, let's say you have a 300-mi trip to make (distance). Because you know that you will go about 60 miles per hour (speed), you estimate that the trip will take about 5 h.

Common units of speed are miles per hour (mi/h), feet per second (ft/s), meters per second (m/s), and knots. A *knot* is 1 nautical mile per hour, or 1.15 statute miles per hour; the speed of ships and airplanes often is measured in knots.

Speed coupled with a direction is *velocity*; therefore, whether you are traveling at 60 mi/h due west or 60 mi/h due north is an important piece of information. These same two speeds (but with different velocities) will land you in very different final locations. Because the direction of motion is typically an important piece of information, problems in physics generally involve velocities, which are represented by a vector. A *vector* is simply any measured quantity that has both a direction and a magnitude, or size.

Velocity is a vector quantity, and speed is a scalar quantity. A *scalar* quantity is one that has a magnitude, but no direction. Mass, or the amount of matter you contain, is another scalar quantity. Velocity, then, can be defined as a change in distance (in a given direction) divided by a change in time. The Greek letter delta (Δ) is used to represent the concept of "change." Therefore, the equation

$$Velocity = \frac{\Delta x}{\Delta t}$$

is read as "velocity equals the change in x divided by the change in t."

As a simple example, if you traveled a distance of 300 mi in 6 h, then your average velocity was 300 mi per 6 h, or 50 mi/h.

The typical unit of velocity in physics problems is meters per second (m/s), but any unit of distance divided by a unit of time in a specified direction represents a velocity. To determine how far an object has traveled (x) from an initial position (x_0), after a set amount of time (t), traveling at constant velocity (v), you would use the following formula (derived from the preceding equation):

$$x = x_0 + vt$$

In the preceding example, if you were traveling with an average velocity of 50 mi/h, and you drove for 10 h on your second travel day, and your starting point (x_0) was 300 mi beyond where you were the day before, then your total distance traveled (at the end of the second day) would be

$$x = (300 \text{ mi}) + (50 \text{ mi/h}) \times (10 \text{ h})$$

$$x = 800 \text{ mi}$$

In problems involving the velocity of an object, there are sometimes several velocities involved, and you will be asked to determine the velocity that results from the sum of all the velocities described. The *resultant velocity* is simply the sum of all the velocity vectors. To determine the *resultant* of a number of vectors graphically, simply place all the vectors "head to tail" (presumably drawn to scale, where the length of the vector is proportional to the velocity). Connecting the first arrow tail to the final arrowhead will show the magnitude and direction of the resultant vector.

As an example, imagine a canoe crossing a river. The movement of the water in the river gives the canoe a downstream velocity, and the person paddling in the canoe gives the boat a velocity across the river (see Figure 3.2). The resultant velocity of the canoe is the sum of these two velocity vectors. The resultant can be determined graphically (as just described) or by using some simple geometry. The lengths of the sides of a right triangle are related by the Pythagorean theorem (see Figure 3.3), which states that in a right triangle the square of the length of the hypotenuse (the long side) is equal to the sum of the squares of the other two sides. Generally, the Pythagorean theorem is written as $a^2 + b^2 = c^2$.

In our example of the canoe crossing the river, the two velocity vectors form a right triangle, so that the resultant velocity can be computed with the formula

$$v \text{ (total)} = (v_1^2 + v_2^2)^{1/2}$$

where v_1 is the velocity of the river, and v_2 is the velocity of the canoe.

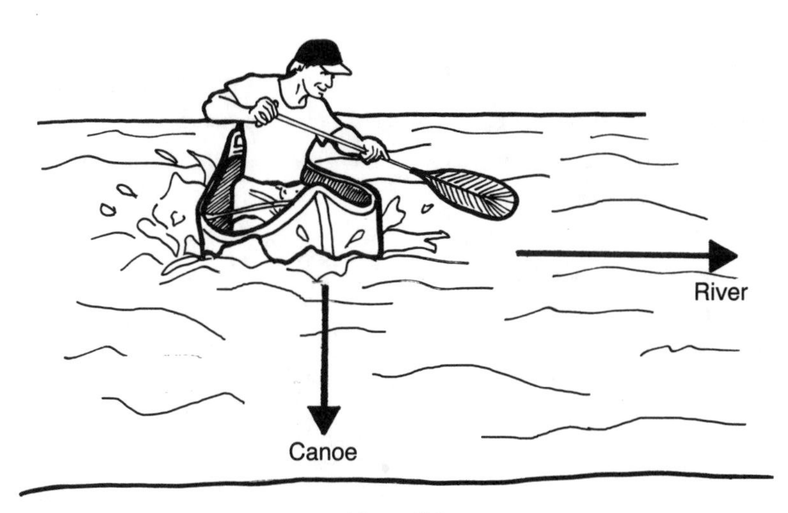

Figure 3.2

Placing the two vectors representing the velocities of the canoe and the river head-to-tail and then drawing a new velocity vector from the tail of the first velocity to the head of the last one produces (graphically) the resultant vector.

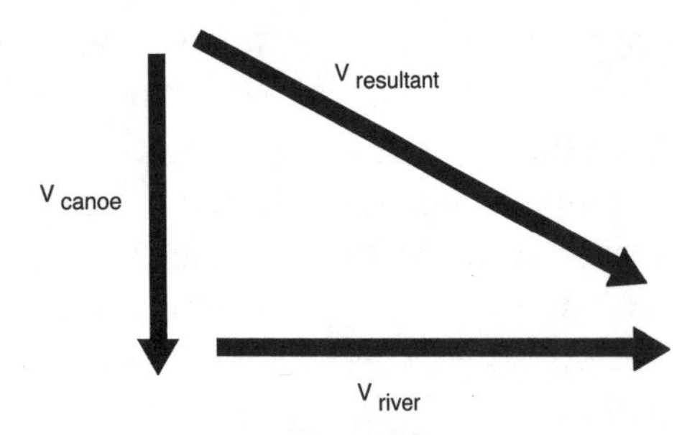

Figure 3.3
The lengths of the three sides of a right triangle are related according to the Pythagorean theorem.

EXPLORATION 3.1

If a river is flowing with a velocity of 4 mi/h downstream, and a girl in a canoe paddles across the river (perpendicular to the river flow) at 3 mi/h, what is the magnitude and direction of the resultant velocity?

Answer:
The resultant velocity is

$$v_{total} = [(3 \text{ mi/h})^2 + (4 \text{ mi/h})^2]^{1/2}$$
$$v_{total} = 5 \text{ mi/h}$$

Acceleration

So far, we have considered only motions involving constant velocity. In the real world, of course, velocities are always varying and are rarely constant. Acceleration is defined as the rate of change of the velocity; that is, the change in velocity divided by the time it takes for the change to occur, or

$$Acceleration = \frac{\Delta v}{\Delta t}$$

Remember that the Greek letter delta (Δ) stands for "change in" the quantity that it precedes. Because velocity is measured in meters per second (m/s) and time is measured in seconds (s), the typical unit of acceleration is meters per second per second (m/s^2).

Imagine that you are sitting at a red light. The light turns green. You press on the gas pedal (the accelerator) and the car moves from a velocity of 0 mi/h to a velocity of 35 mi/h. This change in velocity is called *acceleration,* and the rate at which your velocity changes tells you how great your acceleration is. Car buyers often pay a premium for engines that can take them from 0 to 60 in 6 s instead of 10 s. Why? They are paying for greater acceleration: they are paying to get from 0 to 60 mi/h in a shorter period of time.

Accelerations determine the final velocity of an object. To determine an object's final velocity (v), given that it started at some initial velocity (v_0) and experienced acceleration (a) over a period of time (t), use the equation

$$v = v_0 + at$$

This equation, again, can be derived from the definition of acceleration.

For example, if you are in a car that starts at rest (with an initial velocity $v_0 = 0$) and you accelerate for 5 s at an acceleration rate of 10 m/s^2, then what will be your final velocity? Using the equation

$$v = v_0 + at$$

you see that you will end up (5 s later) at a velocity of

$$v = 0 \text{ m/s} + (10 \text{ m/s}^2) \times (5 \text{ s})$$

or

$$v = 50 \text{ m/s}$$

PHYSICS IN THE REAL WORLD

The crack of the bat indicates that the batter really got a piece of the ball. The right fielder sees the ball coming his way and, from the ball's path through the air, realizes that he is too far out in the field. Instinctively adjusting his speed and direction (his velocity), the player runs toward the ball, pacing himself to arrive at the final location of the baseball at the moment that it is at the height of his glove.

There are two motion problems here, the motion of the batter (running in a straight line), and the motion of the ball traveling along a curve, its path modified by air resistance, which slightly changes the baseball's motion through the air. The brain of the right fielder, as his eyes watch the ball's trajectory, makes subtle corrections to his velocity (accelerations), and as the ball falls into the glove of the running man, the crowd goes wild.

Before we move on, it is important to note that a change in *direction* (not just a change in speed) also constitutes acceleration. Why is this? Well, our definition of velocity is a speed in a given direction. Therefore, a change in direction is a change in velocity, and any change in velocity is acceleration (as we have just defined it).

Thus, when you leave the freeway and enter a curved off-ramp, even if you are moving at a constant speed of 50 mi/h, you are accelerating. This should not come as a surprise, as you can usually "feel" acceleration. Whether you are feeling your car accelerate from 0 to 60, or feeling it as you round a curve, you are feeling the acceleration.

Let's think just a bit more about the example of our accelerating car. Let's say that after 1 s it is going 10 mi/h. Its acceleration is then 10 mi/h/s—not a unit that you use every day but still a proper unit of acceleration (a unit of velocity divided by a unit of time). What if the car is traveling 20 mi/h after 2 s? The acceleration is still 10 mi/h/s, since its velocity (20 mi/h) divided by the time required to reach that velocity (2 s) is 20/2 or 10 mi/h/s; and if the car is going 60 mi/h after 6 s, the acceleration is still 10 mi/h/s. This would be an expensive car.

Slamming on the brakes would result in another kind of acceleration (a negative acceleration, or slowing down, is typically called *deceleration*). For example, suppose a car traveling at 10 m/s is brought to rest by its brakes at the uniform rate of 2.5 m/s^2. How long will it take for the car to stop? If we know that the braking acceleration is -2.5 m/s^2, then the car will slow by 2.5 m/s every second. Thus, a car moving at 10 m/s will be moving at 7.5 m/s after 1 s, and 5.0 m/s after 2 s, and will be at a standstill after 4 s.

PHYSICS IN THE REAL WORLD

You exit the freeway onto a circular cloverleaf that will bring you to an access road. Although you are moving at a constant velocity, your direction is changing at every moment. Because your direction is changing, you are accelerating. What keeps you moving on that curved path? It is your tires. Your tires are in contact with the road, and it is the force between them known as friction that keeps your direction changing, keeps you accelerating.

If the road is icy or wet, there is less friction, because at a microscopic level the surface of the road is smoother. As a result, your tires have a harder time keeping you on that curved (accelerating) path, and your car may leave the road and move off in a straight line. Let's hope not!

EXPLORATION 3.2

Examine how far a braking car will travel while stopping. If the car goes from 10 m/s to a full stop, then its change in velocity is 10 m/s. If it decelerates at a rate of 2.5 m/s^2, it will take 4 s to stop.

Answer:
Using the acceleration formula ($x = x_0 + v_0t + \frac{1}{2}at^2$), and input values $x_0 = 0$ m, $v_0 = 10$m/s, $a = -2.5$m/s^2, and $t = 4$ s, we find that the car will stop in a distance of $x = 0\text{ m} + 10\text{ m/s}(4\text{ s}) + \frac{1}{2}(-2.5\text{ m/s}^2)(4\text{s})^2$

or

$x = 40\text{ m} - 20\text{ m}$
$x = 20\text{ m}$

Free Fall: A Particular Acceleration

Galileo carried out a number of experiments that involved dropping objects from high places and carefully measuring the passage of time. He showed by experimentation that objects appear to fall to Earth with a constant acceleration and that this value (9.8 m/s^2 or 32 ft/s^2) becomes more obvious as one reduces complicating effects like air resistance. The acceleration of gravity at the surface of Earth is often referred to as 1g. Larger accelerations can be referred to as 2g, 3g, and so on. Any object that is falling to the surface of Earth owing to the acceleration of gravity is in *free fall.*

When objects are dropped from a great height, they will eventually stop accelerating and achieve what is called *terminal velocity*. This is the velocity at which the object will not accelerate any more. A leaf has a much smaller terminal velocity than a bowling ball, for example. A feather has a smaller terminal velocity than a hammer, and a flat piece of paper has a smaller terminal velocity than the same piece of paper crumpled up.

Parachutes are a perfect example of the usefulness of lowering one's terminal velocity. A parachute adds almost nothing to a person's mass, but it tremendously increases his or her surface area. A typical parachute has an area that is close to 100 times that of a human, and from the preceding examples, it seems that the key to lowering terminal velocity is a high ratio of surface area to mass. The piece of paper is the best example. The same mass object (a piece of paper) will have a much lower terminal velocity when we allow it to have the greatest possible area (flat).

What Galileo had measured was something called the *acceleration of gravity* (often abbreviated with the letter g). It is a constant value only at the surface of Earth, and only for the particular mass and radius of our planet; however, if we make a few assumptions (that we are close to the surface of Earth and that air resistance is not a significant factor), we can determine a lot about the time it takes for an object to fall to Earth simply by knowing this value of acceleration. The general equation that can be used to determine the position of an object (x) after a time (t) that has a known initial position (x_0), initial velocity (v_0), and constant acceleration (a) is

$$x = x_0 + v_0t + \frac{1}{2}at^2$$

EXPLORATION 3.3

We can use the acceleration equation to solve a simple problem in free fall. Let's say that you are standing at the top of a three-story building, and you want to know how long it will take for the water balloon in your hand to hit the sidewalk below you. Ignoring air friction, the only acceleration involved is the acceleration of gravity (*g*), and the height of the building is 12 m.

Answer:
In the acceleration equation, we use $x = 12$ m (when the balloon hits the sidewalk), $x_0 = 0$ m (we take the balloon's starting point at the top of the three-story building to be zero), $v_0 = 0$ m/s (the balloon starts from rest), and $g = 9.8$ m/s^2 (only gravity is acting on the balloon). Inserting these values into the equation and solving for *t*, we determine that

$$x = x_0 + v_0 t + \tfrac{1}{2} at^2$$

or

$$12 \text{ m} = 0 \text{ m} + 0 \text{ m/s}(t) + \tfrac{1}{2}(9.8 \text{ m/s})t^2$$

or

$$24/9.8 \text{ s}^2 = t^2$$

$$t = 1.6 \text{ s}$$

Thus, for example, if an object is dropped from a height of 100 m and the only acceleration acting is that of gravity (9.8 m/s^2), then we can say that the initial velocity (v_0) is 0 m/s, and the initial height (x_0) is 100 m. We must take a moment to consider direction. If height is measured up (positive) from the ground, then the acceleration of gravity (pointing in the opposite direction) must be negative, or –9.8 m/s^2. Using all this information, we have

$$0 = 100 \text{ m} + 0 \text{ m/s}\,(t) + \tfrac{1}{2}\,(-\,9.8 \text{ m/s}^2)\,t^2$$

or

$$100 \text{ m} = (4.9 \text{ m/s}^2)t^2$$

$$100 \text{ m}/(4.9 \text{ m/s}^2) = t^2$$

$$t = 4.5 \text{ s}$$

so the object takes 4.5 s to strike the ground.

MOTION IN TWO DIMENSIONS

Thus far, although we described position in two and three dimensions, we have been discussing motions in only one dimension and in one direction. Of course, in the real world, we are often interested in the motion of objects in two dimensions (and three dimensions for that matter): the arc that a baseball travels after being struck with a bat from our example earlier in the chapter, or the trajectory of the human cannonball when she is shot from the cannon at the circus.

The key to understanding motion in two dimensions is to realize that an object's motion can be broken into two *perpendicular* components, or components that are separated in direction by 90°. For example, motions in two dimensions can be separated into a component perpendicular to the ground (the *y*-direction) and one parallel to the ground (the *x*-direction). If we treat these two components of velocity separately, we can easily

Figure 3.4
The human cannonball.

understand two-dimensional motion (see Figures 3.4 and 3.5).

Let's think more about the human cannonball just mentioned. Once she is shot out of the cannon at some initial velocity, the only acceleration affecting her (ignoring air friction again) is the acceleration of gravity (g), and gravity acts in one direction: downward perpendicular to the ground. (Actually, gravity acts toward the center of Earth, but locally this looks like perpendicular to the ground.) If we ignore air currents, there is no force acting parallel to the ground. Thus, if we divide the velocity vector of the human cannonball into two components (x- and y-components) the problem becomes much simpler to understand (see Figure 3.5).

The x-component of her velocity (parallel to the ground) is unchanged during her flight. The y-component of her velocity changes gradually, starting with a positive (up) value, slowly changing to zero (as the acceleration of gravity slows her down), and then changing to a negative (down) value (as the acceleration of gravity speeds her up). The combination of these x- and y-direction motions produces a shape known as a *parabola*. A *parabola* is the shape defined in the x-y plane by the equation $y = x^2$. Objects (like baseballs and human cannonballs) that move through the air follow a parabolic *trajectory*, or path.

Let's assume that the human cannonball is shot from her cannon at a 30° angle to Earth and that she leaves the cannon at a velocity of 5.0 m/s. What are her vertical and horizontal velocity components? Figure 3.5 shows a triangle that represents the vertical and horizontal components of her velocity, and to answer the question posed, we need to remember either the Pythagorean theorem (discussed earlier) or a little bit of simple trigonometry.

Her vertical velocity component is equal to her initial velocity times the sine of her angle to the surface of Earth (5.0 sin 30° m/s), and her horizontal velocity component is equal to her initial velocity times the cosine of her angle

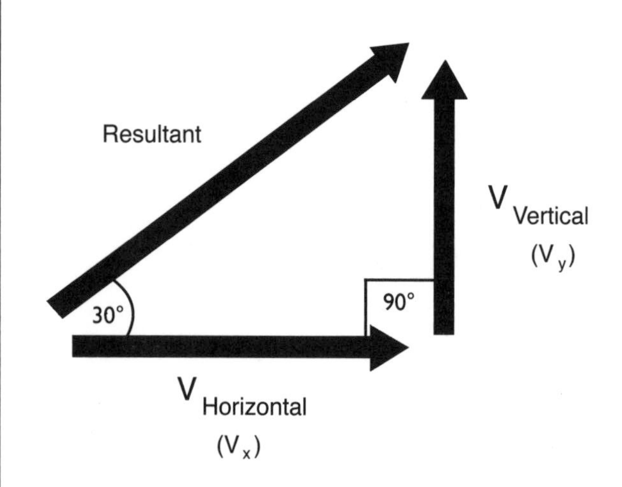

Figure 3.5
Figure shows the vertical and horizontal components of the initial velocity of the human cannonball.

to the surface of Earth (5.0 cos 30° m/s). Thus, her horizontal (v_x) component is 4.3 m/s, and her vertical (v_y) component is 2.5 m/s. You will notice that you can verify these rounded values with the Pythagorean theorem:

$$a^2 + b^2 = c^2$$

or

$$2.5^2 + 4.3^2 = 5.0^2$$

The human cannonball will sail through the air in the horizontal direction at 4.3 m/s until she hits the safety net. Her vertical velocity will change with time, first slowing to zero and then increasing again, because the acceleration of gravity will continuously change the value of her vertical velocity. For more on two-dimensional motion, see Appendix B.

EXPLORATION 3.4

Take a stopwatch and five or six pennies to a window on an upper story of your house. Making sure no one is below you, measure the number of seconds it takes for each penny to hit the ground. Calculate the average of those times. You should be able to use the average time and the height of the fall to estimate the acceleration of gravity (the accepted value is 9.8 m/s²). Use the equation involving distance, velocity, and acceleration, with x_0 being the distance of the fall in meters, $x = 0$ m (end of the fall), $v_0 = 0$ m/s (you drop the pennies from rest), and t being your measured time. Then solve for a (acceleration). What could be some of the reasons why you might get a value different from the accepted value?

PROBLEMS

3.1 You use an Internet map site to calculate the driving distance from your house to your parents' house for the holidays. The results say that the distance is 250 mi, and the travel time is 5 h. What speed is the program assuming you will average for the trip?

3.2 A skydiver reaches a terminal velocity of 10 mi/h at a height of 1 mi above Earth. How long will it take her to reach the ground in minutes?

3.3 A plane flies due south at 250 mi/h, in a 60 mi/h crosswind, blowing from the west. What is the resultant velocity (speed and direction) of the plane?

3.4 A spaceship accelerates from rest at 1g (9.8 m/s²) for 10 min. What is the final velocity of the spaceship after this 10-min acceleration in km/s? In km/h?

3.5 A child accidentally drops her doll from a bridge suspended 50 m above a fast-moving river. Assuming the doll never reaches terminal velocity, how many seconds will it take for the doll to hit the water?

3.6 Two quarters are stacked at the edge of a table. A student uses the blade of a butter knife to knock the bottom coin out away from the table. The top quarter should fall basically straight down. If the table is 1 m high, how long will it take for the top quarter to hit the ground? How long will it take for the other quarter to hit the ground? If you have a couple of quarters and a butter knife, you can try this experiment at home.

CHAPTER 4

FORCE AND MOTION

KEY TERMS

force, resultant force, equilibrium, Newton's laws of motion

WHAT IS FORCE?

In the simplest sense, a *force* is a push or a pull exerted on an object. When you go to the gym and lift weights, you are exerting forces. When you run across the parking lot in the rain to your car, your legs are pushing on the ground, and the ground returns the favor by pushing you forward. As mentioned in Chapter 3 (Objects in Motion), velocity is a quantity that we represent with a vector, because it has both a size and a direction. Forces are also vectors, and like velocities, forces have both magnitude (size) and direction. A downward force is fundamentally different from an upward force. A downward force keeps you anchored to the ground. An upward force can accelerate you into space!

In Chapter 3 we discussed acceleration as a rate of change in velocity. When you push on the accelerator pedal in your car, you cause the car you are in to move at ever-higher velocities. One can observe that objects accelerate—that a dropped object will move at increasing velocity until it reaches its terminal velocity or strikes the ground, whichever comes first; but a question that occurred to scientists in the seventeenth and eighteenth centuries was: What causes acceleration?

As experiments were starting to show during the time of English scientist and mathematician Sir Isaac Newton (1642–1727), in the absence of forces, an object will keep moving at constant velocity in a straight line forever; however, to get an object moving in the first place or to make a moving object move faster or slower requires something else: the presence of a force.

Sir Isaac Newton thought deeply about this phenomenon and other questions. His genius lay in putting together all the pieces that we have discussed so far in this book. In this chapter, we introduce and discuss what have become known as Newton's laws of motion: "rules" of nature that have been subjected to many tests and appear to accurately model (and predict) how velocity, mass, force, and acceleration are interrelated. In physics, forces are generally represented in units of newtons (abbreviated N). A *newton* is the force required to accelerate a 1-kg mass at a rate of 1 m/s^2. In the English system, forces are represented in the more familiar pounds (lb), and 1 lb is the equivalent of 4.448 N. But wait, you say, my weight is not a force. In fact it is! The quantity that you experience as your weight is a function of two parameters: your mass and the mass of Earth (we ignore for a moment that the radius of Earth also matters; we will come back to this point when we fully discuss gravity).

The gravitational pull of Earth determines your weight. Were you to live on a planet with a smaller gravitational pull, your weight would be reduced, but your mass would stay

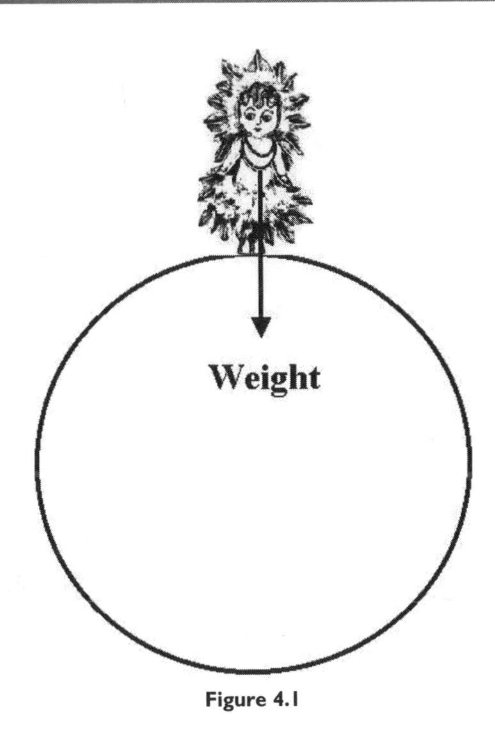

Figure 4.1

the same. Remember, mass is measured in kilograms and is a measure of the quantity of matter present. Your weight is really the measurement of a "pull" on your body exerted by planet Earth. Thus, if you want to lose some weight quickly, go to the Moon. If you want to lose mass, well, that's tougher.

All forces, then, have both a magnitude and a direction and are vector quantities. Your weight is a downward force, with a magnitude (for a given planet) that is proportional to your mass. Vectors are represented with arrows pointing in the direction of the exerted force. Figure 4.1 shows a force vector (representing a person's weight) directed toward the center of the planet, which locally just looks like "down."

More Than One Force

Nature and physics problems are seldom so kind as to have a single force acting on an object. Generally, objects are subject to many forces, and when many forces act on an object we will need to determine their combined effect on the object. In Chapter 3 we discussed vector addition in the context of velocities. Adding vectors works the very same way with

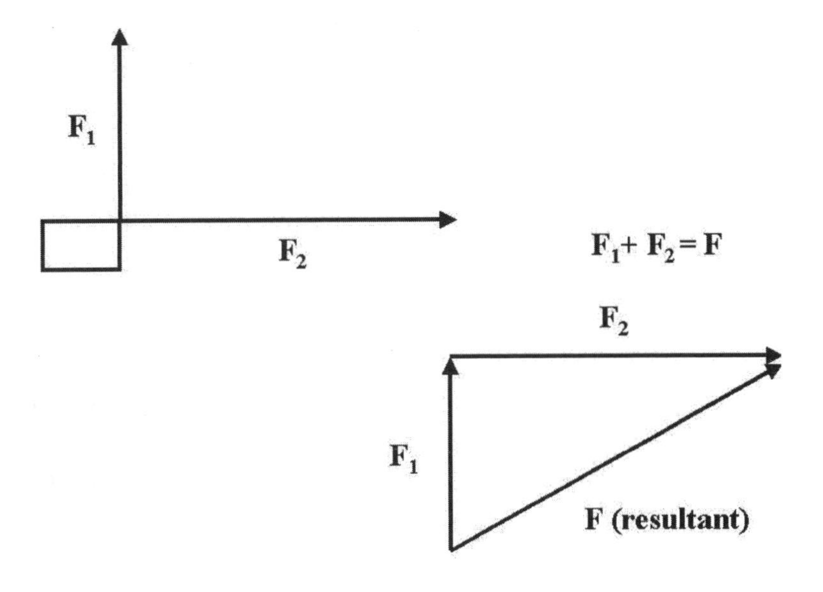

Figure 4.2

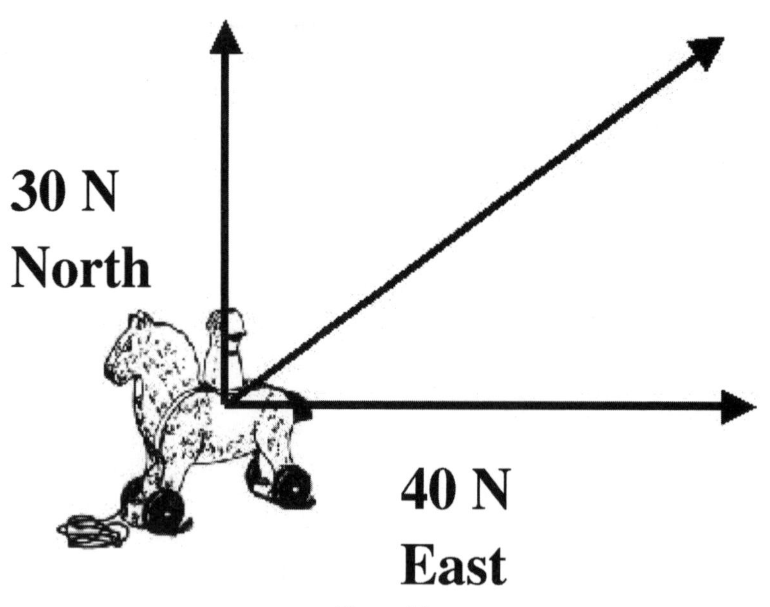

Figure 4.3

forces. When many forces act on an object, physicists add the individual vectors and refer to the sum as a *resultant force*. The resultant force can be determined graphically by placing all the vectors representing the forces acting on an object head to tail and then drawing the *resultant vector* from the tail of the first force to the head of the last one. An example of this process is shown in Figure 4.2.

As a simple example, imagine that there are two teams pulling on a rope in a tug-of-war. One team pulls to the right with a 1000-N force, and the other team pulls to the left with a 1050-N force. The resultant force (easy to determine when they act along the same line) is 1050 N left minus 1000 N right, or 50 N left. Everyone in the tug-of-war will feel the resultant force (sometimes referred to as the *net force*) to the left, and the rope (and people attached to it) will move in that direction.

For a slightly more complicated example, imagine that the forces act perpendicular to one another. Imagine that two children are pulling on a toy, one to the east and one to the north. We will assume that the toy stays in one piece. One child pulls due north with a 30-N force, and the other pulls due east with a 40-N force (see Figure 4.3). We can use the Pythagorean theorem ($a^2 + b^2 = c^2$) to determine that the magnitude or size of the net force is 50 N, but what about the direction? Geometry can help us here. In a triangle with sides of 30-40-50, the smallest angle will be 37° (see Figure 4.3). As a result, the resultant force will be 50 N, 37° north of east (notice that an answer has to specify both a magnitude and a direction).

Forces in Equilibrium

In the example of the two girls and the toy, there is a nonzero resultant force. As a result, the toy will start to move in that direction. Imagine that there is a third girl in the picture, pulling in another direction, 53° south of west, and with a magnitude of 50 N. The three girls (in this scenario) will not move, and as long as the toy does not tear to

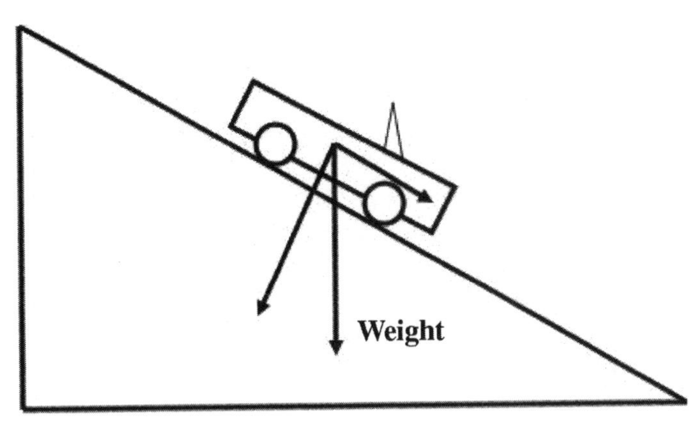

Figure 4.4

shreds, the three girls will have achieved some sort of balance.

When, as just described, all the forces acting on an object add (as vectors) to zero, then we say that the forces acting are in *equilibrium*. Another way to describe such a case is that the net force is zero. In many of the preceding examples we had a resultant force with a nonzero magnitude. What does that mean? Well, if there is a "net force" acting on an object, it will begin to accelerate in the direction of the net force. Therefore, if an object falls off a tall building, the force of gravity will inexorably pull it toward the surface of Earth, until the surface of Earth exerts another force that decelerates the object rapidly. If you forget to apply your parking brake, there is a net force that acts parallel to the surface of the road that will pull your car in that direction (see Figure 4.4).

There are many situations in which we do not want systems to accelerate, though, and in those cases the net force acting on the system must be zero. A building is a good example of this requirement. When a building or bridge is designed, it is important that the net force acting on the structure in question be zero. That result ensures that the building will stand. If you are sitting in a chair right now, the forces acting on you are in equilibrium. Gravity is pulling you down, and your beanbag chair (sitting on the floor beneath you) is pushing you up. These forces are in exact balance, equal and opposite, and you sit there at rest reading these words.

Or imagine that you are an archer in a competition. You place an arrow against the bowstring and pull the bowstring back. Let's say that you pull back with a force of 200 N. As you hold the bowstring there, pulled back, the force that you are exerting must be perfectly balanced by the forces being exerted in two directions by the bow string itself, attached at two points to the bow (see Figure 4.5).
If the vectors representing these three forces were placed head-to-tail as described previously, they would form a closed triangle, with the sum of the three forces being zero. When the bowstring is held in place, forces are exerted in perfect balance, and no accelerations occur. Now, imagine that you release the bowstring. One of the three forces is now removed, and the other two forces are no longer in equilibrium. There is a net force away from you, toward the target. This nonzero net force accelerates the arrow out of your bow in a straight line. Once the

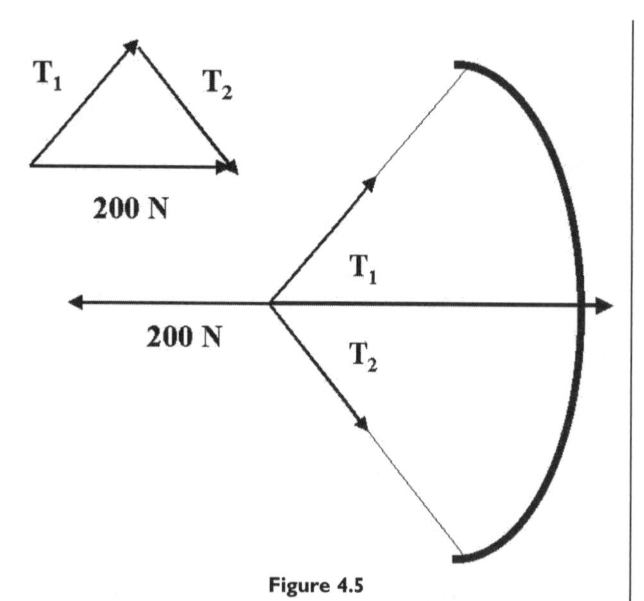

Figure 4.5

bowstring is no longer in contact with the arrow, the acceleration of gravity slowly pulls the arrow (in a parabolic curve) toward the ground. The acceleration of the arrow stops once the force of the bowstring is removed, and the arrow moves at (nearly) constant velocity through the air.

Let's call the two forces in the bowstring T_1 and T_2. If the arrow is centered on the bowstring, the angles between the bowstring and the arrow will be equal, and the magnitude of the two forces (T_1 and T_2) will be equal. If we draw the forces exerted to scale (see inset, Figure 4.5), we can determine the magnitude of the two forces easily.

EXPLORATION 4.1

Using a ruler, determine the tension in each of the bowstrings in the drawing of the archer in Figure 4.5.

NEWTON'S LAWS OF MOTION

Sir Isaac Newton was a man of prodigious talent and with extensive interests, especially in the areas of motion, optics, and gravity. In this chapter we describe his fundamental contributions to our understanding of motion. Through careful observation and modeling, Newton was able to describe the motion of objects by three relatively simple statements, now called *Newton's laws of motion*. These laws include the *law of inertia* (Newton's first law of motion), the *law of constant acceleration* (Newton's second law of motion), and the *law of momentum* (Newton's third law of motion).

Newton's First Law

Newton's first law states something that is not obvious in the everyday world: *all objects remain in their state of rest, or in motion in a straight line, unless acted upon by an outside force*. The first law is often called the law of inertia, because *inertia* is the property of matter that resists changes in its state of motion. The "at rest" part may make more sense at first glance. If you run out of gas, you know that it will take a large exertion of force to get your car (at rest) moving. Moving objects, however, do not keep moving in a straight line forever. When you toss a ball in the air, it does not move in a straight line forever; it falls to Earth. Exactly. The ball falls in a curved path to Earth because it is being acted on by outside forces (the forces of gravity and air friction); but in the absence of gravity and air friction, the ball you tossed *would* move off in a straight line at constant velocity forever. Newton's first law states something rather surprising: that motion at constant velocity is just as natural a condition as rest.

Spacecraft take advantage of this law by firing their engines to get away from Earth and then basically coasting until they get within the gravitational tug of another planet. They move off into space at constant velocity toward their final destination. In some sense, the first law can be thought of as describing objects that are in equilibrium (as described earlier). The net forces acting on them are zero, whether they are at rest or moving in a straight line at constant velocity. Why do they have to be moving in a straight line? Recall that if an object is moving in a curved path, its direction is changing, so its velocity is changing, and it is accelerating; and if it is accelerating, it is not in equilibrium.

PHYSICS IN THE REAL WORLD

You and a friend climb into the car of an old-fashioned, wooden-frame roller coaster ride. The ride begins, and you are rapidly accelerated from rest to a velocity of 10 mi/h. As the car you are in accelerates, the mass of your body tries to resist the motion, so you feel "pushed back" into your seat.
The first turn is a sharp left (remember that a change in direction represents an acceleration). Your body's mass wants it to keep moving in a straight line, so you feel "pushed" into the right side of the car. After a number of other turns, the track drops away suddenly, and again, the tendency of your body to move in a straight line makes you feel as if you are "lifted" out of your seat.

The car slides into the starting area and comes to an abrupt stop, throwing you forward in your seat, your body's inertia again carrying you forward. Your stomach catches up with you a few minutes after the ride.

Newton's Second Law

Newton's second law can most simply be stated as: *force is equal to the product of mass and acceleration, or* ***F*** *=* ***m*****a**. The boldface terms force (**F**) and acceleration (**a**) indicate that, quite properly, these are both vector quantities; that is, they have a direction.

As we have mentioned in several contexts, when the forces acting on an object are not in equilibrium, there is a nonzero resultant vector, and a net force acts on the object. When a net force is acting on an object, Newton's second law tells us that an object acted upon by a constant force will move with constant acceleration in the direction of the force. The force and the acceleration are vector quantities, and they must therefore act in the same direction. The magnitude of the acceleration will be directly proportional to the acting force and inversely proportional to the mass of the body. Because mass is measured in kilograms (kg) and acceleration is measured in meters per second (m/s^2), the basic unit of force is the product of these units, or $kg{\cdot}m/s^2$, the newton (N).

Have you ever heard a reference to *g*-forces? These are forces that are exerted on pilots and astronauts when they are accelerated at rates higher than the acceleration of gravity. Thus, if you are accelerated at 1 *g*, then you are accelerating at 9.8 m/s^2 (or 32 ft/s^2), the normal value at the surface of Earth. An acceleration of 2*g* is simply double this value, or 19.6 m/s^2 (64 ft/s^2). Being accelerated at rates higher than normal for the surface of Earth will make you feel "heavier," as if someone suddenly doubled the mass of Earth, and your weight doubled. A pilot often feels these accelerations when his aircraft makes a turn. You probably have felt forces like these when you have been on a roller coaster ride or taking off in an airplane. You will feel these *g*-forces in any situation in which you are accelerating rapidly.

EXPLORATION 4.2

Your weight is actually a measure of the force of gravity acting on the mass of your body. For example, let's say that you want to find your mass in kilograms (kg) using Newton's second law. Use your weight in pounds (lb) and the acceleration of gravity ($g = 9.8$ m/s^2) to determine your mass in kilograms. (*Hint:* Use the conversion factor 1 lb = 4.448 N.)

Newton's Third Law

Newton's third law states that *for each action there is an equal and opposite reaction*. What does this mean? In the case of your hammering a nail, it means that the hammer exerts a force on the nail (driving it into the wood), and the nail exerts an equal and opposite force on the hammer, stopping its forward motion. In the case of your running down the street it means that your feet push backward on the street, and the street pushes forward on you, pushing you forward. The street (and the earth that it is attached to) is much more massive than you, so your forward motion is much more appreciable than the tiny back-

Figure 4.6
The launch of the ROSAT X-ray telescope on June 1, 1990, aboard a DELTA II rocket. (Image courtesy of NASA.)

ward motion of Earth; however, the forces exerted are equal and opposite. Newton's second law ($F = ma$) can help our understanding here. The forces exerted by you and the planet are equal, but your mass is much smaller; therefore, the acceleration that you experience is far greater.

A spaceship moves forward because it expels combusted fuel backward. The forces exerted are equal. The spaceship exerts a backward force on the combusted fuel, and the combusted fuel exerts a forward (opposite in direction) force on the spacecraft (see Figure 4.6).

Newton's laws of motion apply to an enormous array of phenomena, from balls to airplanes to rocket ships. We will revisit Newton's laws many times in coming chapters.

EXPLORATION 4.3

You point a gun at a monkey hanging in a tree. (Note to animal rights activists: this particular monkey is infected with a deadly virus and must be destroyed.) In order to hit the monkey—if it lets go of the branch at the moment you pull the trigger—should you point the gun *at* the monkey, *above* the monkey, or *below* the monkey? See Figure 4.5.

Answer:
Both the monkey and the bullet are subject to the same accelerating force (gravity); therefore, you should point the gun directly at the diseased monkey. In the absence of gravity, the gun pointed at the monkey would deliver the bullet properly. In the presence of a gravitational field, both the monkey and the bullet move downward the same distance in each instant of time, so the bullet will find its mark.

PROBLEMS

4.1 Like velocity, force has both a(n) ——— and a(n) ———.

4.2 Newton's first law of motion refers to an object's ———.

4.3 Newton's second law of motion gives the relationship among three quantities: ———, ———, and ———.

4.4 If the archer described in this chapter pulls back with a force of 300 N, and the angle between the arrow and the bowstring is 45°, what will be the tension on each side of the bowstring? What part of the archer is providing that forward force?

4.5 A truck is parked in the middle of a 100-m-long bridge. If the truck has a mass of 1000 kg, what is the upward force exerted at each end of the bridge? Does the length of the bridge affect your answer?

4.6 What force is required to hold a 100-kg safe motionless at the surface of Earth? What force is required to move the safe upward with constant velocity? What force is required to accelerate the safe upward at 1 m/s^2? What force is required to accelerate it downward at 9.8 m/s^2?

CHAPTER 5 WORK, ENERGY, AND COLLISIONS

KEY TERMS

work, energy, potential energy, kinetic energy, conservation of energy, momentum

WORK AND FORCE

How do you use the word *work*? You might say, "Work was rough today," or "Who knew that relationships could be so much work?" Like many words, *work* has a different meaning in physics than it does in the everyday world. However, what if you walked into your apartment, collapsed on the couch, and said, "It sure took a lot of work to get that dresser into the truck!" You would mean that you had to exert a force to move something. In the last chapter we discussed the nature of forces and how forces result in accelerations that can be described by Newton's second law ($F = ma$). In physics problems, when forces act over some distance, we say that they do *work*. Expressed mathematically, the work done by a force is written as

Work = Force × Distance

or

$W = fd$

From previous chapters, we know that distance is measured in meters (m), and forces are measured in newtons (N); therefore, the unit of work is a newton-meter (N · m), which is sometimes called a joule (J). The joule is a derived unit, because it is derived from the fundamental units that we defined in Chapter 2. A joule (named after English physicist James Prescott Joule) is thus 1 kg · m^2/s^2. One very important fact to remember about work is that the only force that matters in calculating work is the force exerted *in the direction of motion.*

EXPLORATION 5.1

Hercules lifts a large boulder over his head and holds it there threateningly. If he lifts the boulder 2 m in the air and the boulder weighs 1000 N, what amount of work does he do in lifting the boulder? Answer: If we use $w = f\,d$, then the work done is simply 1000 N 2 m, or 2000 N·m, or 2000 J. How much work does Hercules do if he walks 1 m to the left at constant velocity? The answer is none, since the force exerted (up) on the rock is not exerted in the direction of motion.

The entire universe consists primarily of two constituents: matter and energy. Cosmologists—astronomers who study the early universe and the largest scales we can comprehend—often talk about the interplay between matter and energy. Einstein famously stated the relationship as $E = mc^2$: energy is equal to mass times the speed of light squared. We will discuss the exact meaning of this statement in coming chapters; however, for now, because physics is the study of matter and energy, we begin to discuss energy.

We just defined work as the result of a force acting over some distance. *Energy* is often defined as *the ability to do work.* If something has a large amount of energy, it has the ability to do a lot of work. The boulder suspended over the head of Hercules in Exploration 5.1 has the ability to do work. Held over his head like that, the boulder has what is called *potential energy* (abbreviated PE).

Potential Energy

An object's *potential energy (PE)* depends on its position. For example, a boulder held over your head has more potential energy than a boulder resting on the ground. The boulder held at a height of 2 m over your head has an amount of PE that depends on its weight (mass times the acceleration of gravity) and its height; and a stretched spring has more potential energy than an unstretched spring, because when you release it, the spring will return to its normal "resting" position, and in that return the spring can do some work.

There are different types of potential energy. *Gravitational potential energy* is potential energy that is related to moving objects in a gravitational field and is calculated as the product of the object's weight and its height above some defined zero-level. We often write gravitational potential energy (GPE) as

$$\text{GPE} = \text{Weight} \times \text{Height}$$

where weight is mass times the acceleration of gravity, or

$$\text{GPE} = mgh$$

Notice that the unit of gravitational potential energy is newton meter (N · m), the same unit used to measure work. If we do work on an object (by lifting it above our head, for example), we give it an amount of potential energy equal to the work we did!

Chemical potential energy is the stored energy in chemical bonds. The gas tank in your car contains chemical potential energy, because when the gas is burned in the engine the engine can do work, exerting a force to move your car through some distance.

Kinetic Energy

As the name implies, the *kinetic energy* of an object is related to its motion. Moving objects have energy—have the ability to do work—as you know from experience. Why don't you let a baseball hit you in the face? Because of the significant work it would do on your nose; and the faster the baseball is moving, the more work it can do, the more energy it has. The more massive the baseball, the more work it can do, the more energy it has. Specifically, we define the *kinetic energy* (KE) of an object as half its mass times its velocity squared, or

$$\text{KE} = \tfrac{1}{2} mv^2$$

From this formula, it is apparent that the more massive an object, and the faster it is moving, the more kinetic energy it possesses. Thus a slow-moving truck and a fast-moving bullet may have an equivalent amount of kinetic energy, although their masses are very different. The units of kinetic energy are determined by taking the product of the units for mass (kg) and velocity squared (m^2/s^2). Thus the units of KE are (not surprisingly) $kg{\cdot}m^2/s^2$, or joules. The unit of energy is the joule, whether it is kinetic energy or potential energy. A very small unit of energy often used is the electron-volt (eV), which is equal to 1.602×10^{-19} J.

EXPLORATION 5.2

A truck and a bullet are heading toward you. The truck is a 5000-kg truck, moving at 50 km/h, the bullet is a 100-g bullet, moving at 500 m/s. Which object has more momentum? Which one has more kinetic energy?

Answer:
Momentum is mass times velocity, so to compare apples with apples, let's convert all the masses to kilograms and all the velocities to meters per second. First, let's calculate the momenta:

Truck: $p = mv$
$= (5000 \text{ kg})(50 \text{ km/h})(1 \text{ h}/3600 \text{ s})(1000 \text{ m}/1 \text{ km}) = 70{,}000 \text{ kg·m/s}$

Bullet: $p = mv$
$= (100 \text{ g})(1 \text{ kg}/1000 \text{ g})(500 \text{ m/s}) = 50 \text{ kg·m/s}$

Kinetic energy:

Truck: $KE = \frac{1}{2} mv^2$
$= \frac{1}{2}(5000 \text{ kg})[(50 \text{ km/h})(1 \text{ h}/3600 \text{ s})(1000 \text{ m}/1 \text{ km})]^2$
$= 480{,}000 \text{ kg m}^2/\text{s}^2$, or 4.8×10^5 J

Bullet: $KE = \frac{1}{2} mv^2$
$= \frac{1}{2}(0.1 \text{ kg})(500 \text{ m/s})^2$
$= 13{,}000 \text{ kg m}^2/\text{s}^2$, or 1.3×10^4 J

Thus although the truck has over 1000 times more momentum than the bullet, it has only about 50 times more kinetic energy.

EBB AND FLOW AND THE CONSERVATION OF ENERGY

If we ignore effects like friction and air resistance for the time being, we can say that the total energy of a system, represented by the sum of its kinetic and potential energy, is conserved. Mathematically, we can express the law of *conservation of energy* as

Total Energy = Kinetic Energy + Potential Energy

or

$TE = KE + PE$

Imagine you are holding two cups, one full of water and one empty. Label the full cup KE and the other cup PE. Now, pour half of the water from the cup labeled KE to the cup labeled PE. The total amount of water that you have in the two cups remains constant, although the amount in each cup may change as you pour water back and forth. You can see that you may have more potential energy or more kinetic energy at any instant, but the total (in an ideal world) does not change.

What happens, though, as you pour the water back and forth? Does a little water spill? If you were to pour the water back and forth many times, your total amount of water would decline slightly. That is more like how a real system works. Over time, energy is "lost" to processes like friction and air resistance.

A child in a swing is a perfect example of this "ebb and flow" between kinetic and potential energy (see Figure 5.1). At the top of the swing, the child hangs motionless for a

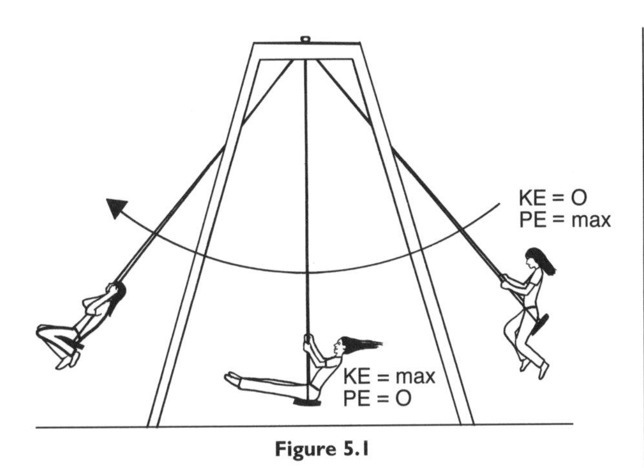

Figure 5.1

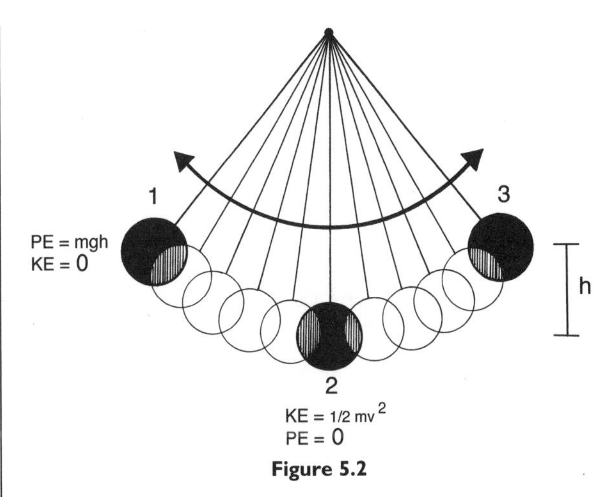

Figure 5.2

moment, possessing only potential energy. Gravity pulls on her and the swing, and the ropes cause her to move in a curved path. At the lowest point in the swing, her energy is purely kinetic (moving) energy. This motion carries her to the other end of the motion of the swing, her kinetic energy decreasing to zero at the top of the swing, where her energy is again all potential. As any parent knows, a child in a swing will not, unfortunately, swing forever. It is necessary to stand behind the child, pushing her at the end of the swing, adding a little bit of energy to the system, energy that is being "lost" to air resistance and other factors. If you did not stand there and push, each arc would become smaller, and the swing would move more and more slowly until it came to rest at the lowest point of the arc.

Figure 5.2 shows a schematic of a pendulum, which has a motion similar to that of a child on a swing. The maximum potential energy comes at the end of the swing (Position 1), where the value of the potential energy (PE) is *mgh*. All this potential energy is converted to kinetic energy at the bottom of the swing (Position 2), and the kinetic energy is $\frac{1}{2} mv^2$, as described previously.

EXPLORATION 5.3

Tie a washer to a 20-cm piece of string. Holding the other end of the string, pull the washer back and allow it to swing back and forth like a pendulum. What is the approximate time it takes for the pendulum to go back and forth once? This time is called the period (*T*). Now, halve the length of the string to 10 cm and allow the pendulum to swing once more. Does one full swing now take more or less time? What happened to the period? What do you think will happen when you tie a 30-cm piece of string to the washer? Test your prediction. Based on this experiment, why, would you say, are swings with long ropes the most fun?

MOMENTUM AND COLLISIONS

An object in motion possesses another attribute called *momentum*. Momentum is often abbreviated with the letter **p** and is equal to the product of mass and velocity, or $\mathbf{p} = m\mathbf{v}$.

The units of momentum are the product of the units for mass and velocity, or kg•m/s. Of course, any appropriate units of mass and velocity may be used, but these are the most common.

Newton's third law (for every action there is an equal and opposite reaction) is sometimes referred to as the *law of conservation of momentum*. This means that in a closed system the total momentum does not change. Notice also that since momentum is the product of a mass (scalar) and a velocity (vector), momentum is a vector quantity, meaning that it has a direction. The direction of the momentum is often very important in problems dealing with conservation of momentum.

Consider the following situation. It is a hot day. You are standing on a motionless boat in your swimsuit. You are part of a system with zero momentum. You decide to jump out of the boat into the water to cool off. You go in one direction at a given velocity, with a certain momentum. You and the boat represent a "closed system," so in order for momentum to be conserved (in order for the total momentum of the system to be zero), the boat has to go in the opposite direction. If the boat is large, like an ocean liner, its velocity in the opposite direction will be small, probably imperceptible. If the boat is small, like a canoe, its velocity away from you will be about the same as your velocity away from it (if your masses are about equal). To restate this situation as a conservation of momentum problem, we write that the total momentum is

$$\mathbf{p}_{\text{total}} = 0 = m_1\mathbf{v}_1 + m_2\mathbf{v}_2$$

or

$$m_1\mathbf{v}_1 = -m_2\mathbf{v}_2$$

indicating that the two objects (the boat: m_1, and the person: m_2) will move off in opposite directions. If the mass of the boat (m_1) is 100 kg and the mass of the person (m_2) is 50 kg, then the velocity of the boat ($\mathbf{v}_1$) will be

$$\mathbf{v}_1 = (-m_2/m_1)\ \mathbf{v}_2$$

or

$$\mathbf{v}_1 = -0.5\ \mathbf{v}_2$$

indicating that the boat will move away from the jumper at half the speed of the jumper moving away from the boat. The reason for calling Newton's third law the law of conservation of momentum should be a little clearer now.

PHYSICS IN THE REAL WORLD

Power (P) is a concept that is important in the everyday world. Power is the rate at which work is done. Doing a large amount of work in a small amount of time requires a lot of power. Power is expressed as

Power = Work/Time

or

$P = W/t$

The most common unit of power is the watt (W), and 1 W is 1 J of work per second; 1000 J in 1 s is 1000 W, or a kilowatt. If you look at your "power bill," you will see that you are billed for a certain number of kilowatt-hours. Thus, the units of your power bill are actually not units of power but units of work and energy (power times time); so it's really an energy bill.

PROBLEMS

5.1 A student is pushing a box of her belongings across a wooden floor. A constant force of 200 N is required, and she pushes the box over a distance of 10 m. How much work does the student do?

5.2 Which has more potential energy, a 500-kg anvil suspended 10 m in the air, or a 3-kg bowling ball suspended at 200 m. How much work was required to get each object from the ground to their higher potential energy positions?

5.3 An electron has a mass of 1.67×10^{-23} kg. If it is accelerated in a particle accelerator to 3% of the speed of light, what is its KE in joules? What is its KE in electron-volts (eV)? The speed of light is 3×10^8 m/s.

5.4 A boy stands on a 50-m-high bridge with his GI Joe™. The action figure has a real parachute. The boy drops Joe from the bridge, but tragically the chute fails to deploy. Foul play? If the action figure has a mass of 100 g and it never reaches terminal velocity, how fast will Joe be going when he strikes the river?

5.5 A 25-kg child is on a swing. If her height is 2 m when she is at the maximum height of her swing, how fast will she be going at the lowest point in her swing? Describe in words any of the assumptions you make in this calculation.

5.6 In the course of a 30-min workout session you lift a total of 10,000 kg, an average of 50 cm. How much work did you do? How much power was required for your workout?

CHAPTER 6

CIRCULAR MOTION AND GRAVITY

KEY TERMS

angular motion, angular velocity, angular acceleration, torque, universal gravitation

The physical world is filled with examples of circular motion: children on a merry-go-round, planets orbiting the Sun, the propeller of an airplane, a yo-yo. Clearly, circular motion, or more generally, *angular motion*, is an important part of the way objects move in the world. In this chapter, we define the terms needed to describe and predict angular motion and outline the parallels to our descriptions of *linear motion* in Chapter 3.

ANGULAR MOTION

When we first described motion in Chapters 2 and 3, we needed to use a unit of distance: we chose the meter. Linear distances are measured in meters and in other linear units of measure, as displacements from an arbitrary zero-point. *Angular motion* is measured somewhat differently. Imagine that you are standing facing north. We will call this direction 0°. Now, rotate your body so that you are facing to the east. You have undergone angular motion; but how "far" have you gone? You have not moved in terms of meters, but you have rotated through an angle of 90°. Your angular motion was 90°. Now, keep turning until you are facing south. Now your angular motion is 180°. Face due west and you have rotated through 270°, and when you are facing north again you have rotated through 360°. Thus, angular motion is measured in degrees (°) or radians (rad), which have no "units" (see Exploration 6.1). Another way to measure angular motion is in revolutions (one full rotation is a revolution). A radian (rad) is an angular measure scaled to the circle. It is the angle for which the arc length on a circle of radius r is equal to the radius of the circle. In one complete revolution, the arc length is the circumference of a circle. Thus there are 2π rad in one complete revolution, so π rad in a half revolution and $\pi/2$ rad in a quarter revolution (see Figure 6.1).

EXPLORATION 6.1

A radian is a unit used to measure a change in angular position. There are 2π rad in a circle. How do we convert from degrees to radians?

Answer:

If there are 2π rad in 360°, then there are π rad in 180°. To convert from degrees to radians, we simply multiply:

$$\text{Radians} = \text{Degrees} \times (\pi/180)$$

To convert from radians to degrees, we multiply:

$$\text{Degrees} = \text{Radians} \times (180/\pi)$$

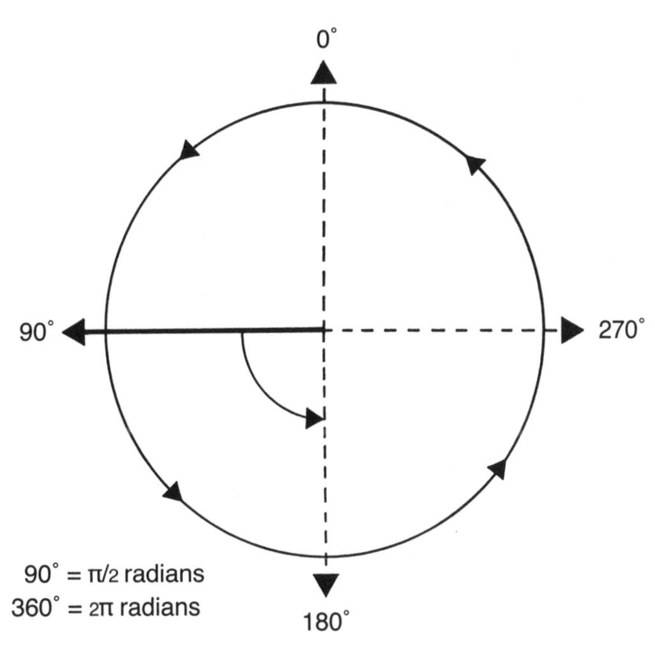

Figure 6.1
Degrees and Radians in Circular Motion

ANGULAR VELOCITY

You may have noticed that your car has a gauge called a *tachometer* that measures the number of revolutions per minute (rev/min) of the engine's crankshaft. You will notice that when the car is not moving that the tachometer will read somewhere around 500 or 1000 rev/min, sometimes known as the *idle speed*. Idle speed? How can you have any speed when you are idling (not moving)? The speed referred to is the *angular speed*, or the number of revolutions per minute that the engine's crankshaft is making. When you press the "accelerator" you feed more fuel to the engine, causing the crankshaft to rotate at a higher angular velocity—whether or not the car is moving. The motion of the crankshaft is transferred to the wheels of the car by the transmission (automatic or manual).

Angular velocity, then, is simply a measure of the rate of change in angular position. In our first example, imagine that you make one full revolution in 10 s. Then, your angular speed is 360° per 10 s, or 36° per second; you could also say 0.1 rev/s, or 0.2π rad/s. All these expressions are equivalent, but the most commonly used units in problems are radians per second and revolutions per minute. Because radians are dimensionless units, the unit of angular velocity is simply 1/s.

PHYSICS IN THE REAL WORLD

The Sun appears to move across the background stars very slowly. This apparent motion occurs, of course, because the Earth orbits the Sun at an angular speed of approximately 2π rad per year, or 360° per year, or 360° in 365.25 days, which works out to a little less than 1° (0.99°) per day. Does this give you any ideas about why there might be 360° in a circle?

ANGULAR ACCELERATION

Of course, if there are angular velocities, then there must be angular accelerations. What does it mean to have an angular acceleration? Picture a stationary merry-go-round. Because it is not rotating, its angular velocity is zero. Now, you start pushing on it so that it starts to turn at a velocity of one revolution every 4 s, or 0.25 rev/s. The merry-go-round has been accelerated to this nonzero angular velocity. Now, if it continues to move at a constant angular velocity, its angular acceleration is zero.

Accelerations are *changes* in velocity, so the units of angular acceleration are radians per second per second. Again, because radians are dimensionless units, angular acceleration is measured in units of $1/s^2$. There are parallels to linear motion in all of these situations. Angular position is represented by the Greek letter theta (θ), angular velocity by the Greek letter omega (ω), and angular acceleration by the Greek letter alpha (α). In the same way that velocity is change in position divided by change in time, angular velocity (ω) is change in angular position divided by change in time, or

$$\omega = \Delta\theta/\Delta t$$

and *angular acceleration* is change in angular velocity over change in time, or

$$\alpha = \Delta\omega/\Delta t$$

Another way to think of angular velocity is as a linear velocity (v) divided by a radius (r), or

$$\omega = v/r$$

We often refer to this speed as the tangential speed (v_t), or

$$v_t = r\omega$$

This relationship should make sense, because if an object is moving at a given tangential velocity (v_t), the larger the radius of its circular path, the smaller its angular velocity.

In a parallel sense, the tangential acceleration (a_t) can be expressed as

$$a_t = r\alpha$$

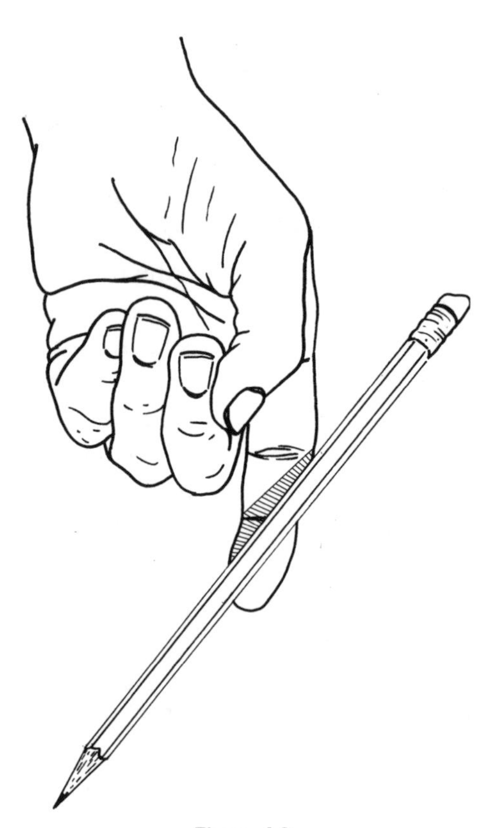

Figure 6.2
The Center of Mass of a Pencil

Torque

The parallel for the linear concept of force is the angular concept of *torque*, represented by the Greek letter *tau* (τ), where *torque* is simply the application of a force at some distance from a pivot point (see Figure 6.3), or

Torque = Force × Radius

or

$\tau = Fr$

The only factor we have not considered yet is mass. Mass is the property of an object that keeps it in its state of rest or motion (recall that Newton's first law is sometimes called the *law of inertia*). In addition to mass, all objects have a property known as their *moment of inertia* (I). This property is a measure of how the mass is distributed and affects how easy or how hard it is to start an object rotating (or stop it from rotating). Thus when we discuss rotational motion, instead of referring to an object's mass, we refer to its moment of inertia (I). The rotational equivalent of Newton's second law, then, states that torque is equal to the product of the moment of inertia and the angular acceleration:

$\tau = I\alpha$

Another way to think of this relationship is that all net torques produce an angular acceleration (in the same way that all net forces produce a linear acceleration). In the case of the merry-go-round, the exertion of a torque (pushing with some force at the edge of the merry-go-round) resulted in the acceleration of the merry-go-round from a state of rest (in an angular sense) to a state of motion. Once the force is removed, the merry-go-round should continue to rotate (in the absence of friction) at constant angular velocity.

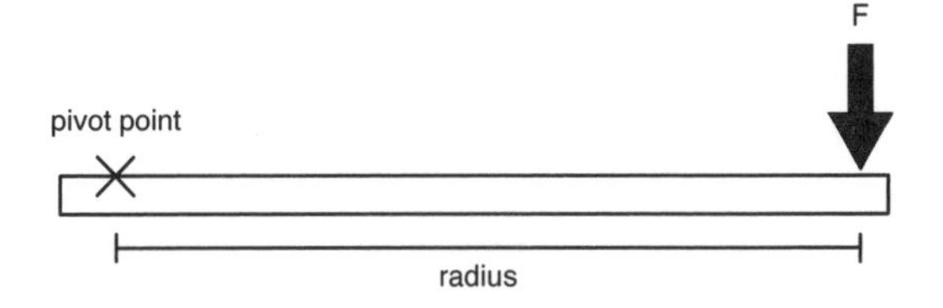

Torque = r F

Figure 6.3

> **EXPLORATION 6.2**
>
> All objects have a point called their *center of gravity* or *center of mass*. This is the point in an object where all its mass can be thought to be concentrated. For example, a pen with a cap on it will have a center of gravity offset from its halfway point, because there is more mass on the end with the cap. If you toss the pen in the air, spinning, it spins not about its physical center, but about its center of mass. Try it! You can find the center of mass of a pencil by trying to balance it (sideways) on your finger (see Figure 6.2). If you place a glob of Silly Putty™ on one end of the pencil, you shift its center of gravity significantly to that end. Now toss the pencil in the air, spinning. What does the pencil appear to do?

ANGULAR MOMENTUM

In the same way that moving objects have linear momentum, rotating objects have angular momentum. The parallel-to-linear momentum ($\mathbf{p} = m\mathbf{v}$) is angular momentum (L), which is equal to the product of the moment of inertia (replacing mass) and angular velocity (replacing velocity):

$$L = I\omega$$

Expressing the preceding equation in terms of the more familiar units of mass, velocity, and radius, we say that a body of mass m moving in a circle of radius r at a velocity v has an angular momentum of

$$L = mvr$$

Conservation of Angular Momentum

Just like linear momentum, angular momentum is conserved, which means that in a closed system angular momentum is neither created nor destroyed. Another way to state this law is that the angular momentum in one state is equal to the angular momentum in another state, and that value is a constant.

Expressing this statement mathematically, we say:

$$L_1 = m_1v_1r_1 = L_2 = m_2v_2r_2 = \text{Constant}$$

The classic example of *conservation of angular momentum* is a spinning ice skater. She starts to spin at a given velocity with her arms outstretched. As she pulls her arms in she is redistributing her mass to a smaller radius. Therefore, since her mass is constant and the radius is decreasing (as she pulls her arms in), conservation of angular momentum tells us that the ice skater must start to spin faster, and of course she does (see Figure 6.4). Table 6.1 compares the parallel terms used to describe linear and angular motions.

As the skater pulls mass (i.e. arms) closer to her axis of rotation, she spins faster, conserving angular momentum

Figure 6.4

> **EXPLORATION 6.3**
>
> If you have an office chair that spins, with your arms outstretched hold a book in each hand and start spinning yourself. Quickly pull the books into your chest. What happens to the speed at which you are spinning? You have just experienced the conservation of angular momentum.

FORCES IN CIRCULAR MOTION

What keeps objects moving in the circular paths that we have been discussing? In some cases, it is a rope or string. There is a yo-yo trick called "Around the World" in which the yo-yo, spinning free at the end of the string, is sent around in a huge circle. Clearly, in

Linear Motion	Units	Angular Motion	Units
Distance(x)	m	Angle (θ)	rad, °
Velocity (v)	m/s	Angular Velocity (ω)	rad/s, or 1/s
Acceleration (a)	m/s^2	Angular acceleration (α)	rad/s^2 or $1/s^2$
Force (F)	N	Torque (τ)	N · m
$F = ma$	N	$\tau = I\alpha$	N · m
Mass (m)	kg	Moment of inertia (I)	$kg \cdot m^2$

Table 6.1 The Parallel Terms for Linear and Angular Motion

this case the string keeps the yo-yo moving in its circular path. Without the string, the yo-yo would move off in a straight line and perhaps strike someone. The tension in the string provides the force that keeps the yo-yo in circular motion.

When you are standing on a merry-go-round, friction keeps you on the surface, and thus friction provides the force to keep you moving in a circle. Imagine standing on an icy merry-go-round—not much friction, and it would be a lot harder to stay on that merry-go-round. It might even be impossible. And if the merry-go-round were to spin fast enough, no amount of friction could keep you on the surface; you would have to try to hang on for dear life!

When you exit an interstate highway, traveling the circumference described by the cloverleaf turn, friction between your tires and the road keeps you moving in that curved path instead of in a straight line. If the roads are icy or wet (and friction is reduced), you may in fact move in a straight line and not a circle as you leave the freeway.

But what about one of the greatest examples of circular motion: the nearly circular orbits of the planets around the Sun? There is no rope or friction to keep the planets in their curved trajectories. There is no connection between the Sun and the planets, but there is a force—the same force that keeps you stuck to the ground and pulls apples out of trees: gravity.

UNIVERSAL GRAVITATION

Johannes Kepler (1571–1630) first proposed, based on the careful observations of Tycho Brahe (1546–1601), that the planets move in elliptical paths around the Sun; but it was Newton who first described the force that makes them move in these paths. Newton made a great intellectual leap at the time, claiming that the same physical laws that can explain how things move here on Earth (balls and apples) can explain the motions of the planets, which are after all, just very big balls. Newton posited that all objects with mass exert an attractive force on all other objects with mass, and that this force is directly proportional to the product of the masses of the two objects and inversely proportional to their distance squared. One of the most pleasing things about the force that Newton proposed was that it predicted (based on theoretical

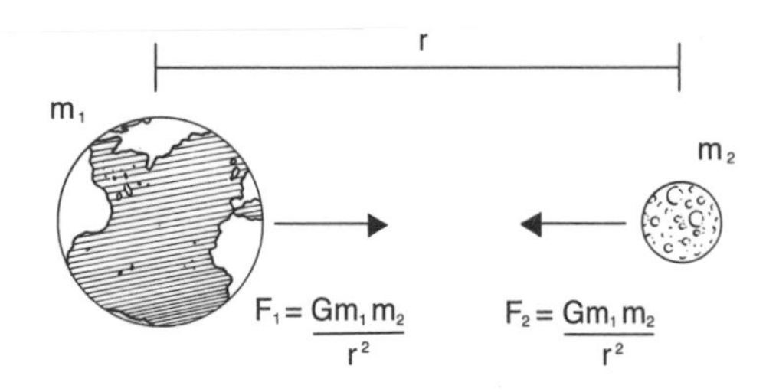

The moon (m_2) and the Earth (m_1) exert equal and opposite forces on one another ($F_1 = F_2$)

Figure 6.5

grounds) that the planets should orbit the Sun in elliptical paths, as the observations of Tycho Brahe and the mathematics of Kepler had shown was in fact the case.

EXPLORATION 6.4

Use Newton's law of universal gravitation (next column) to determine the attractive force between you and Earth. Earth has a radius of about 6.38×10^6 m and a mass of 5.98×10^{24} kg. If your mass is 80 kg (equivalent to about 200 lb), then the force of gravity exerted on you would be

$$F = 6.67 \times 10^{-11}\ \text{N} \cdot \text{m}^2/\text{kg}^2(80\ \text{kg})(5.98 \times 10^{24}\ \text{kg})/(6.38 \times 10^6\ \text{m})^2$$

or 784 N.

You could also determine this force (otherwise known as your weight) by using the fact that weight is equal to mass times the acceleration of gravity (g).

$W = 80$ kg (9.8 m/s^2) or

$W = 784$ N

Newton's law of *universal gravitation* is written:

$$F_G = Gm_1m_2/r^2$$

where m_1 and m_2 are the masses of the two objects, r is the distance between them, and G is the gravitational constant, a very small number that is a measure of the strength of the gravitational force in the universe. The value of G is $6.67 \times 10^{-11}\ \text{N} \cdot \text{m}^2/\text{kg}^2$, and the force of gravity (like all forces) is measured in newtons (see Figure 6.5).

PHYSICS IN THE REAL WORLD

Our planet, Earth, orbits the Sun in an elliptical path that is nearly circular. If we assume for a moment that the orbit is circular, we can simplify a calculation of how fast we are moving (our tangential speed) through the solar system. We are moving at a rate of 360° per year, or 0.99° per day, or in radians (see Exploration 6.1), 2π rad (one complete orbit) per year. There are 3.16×10^7 s in a year, which results in an angular velocity of 1.99×10^{-7} rad/s. So what is our tangential velocity (in the direction of motion)? We use the fact that the average distance between Earth and the Sun (1 A.U.) is 1.5×10^{11} m.

We know that $v_t = r\omega$, so

$$v_t = r\omega, = (1.5 \times 10^{11}\ \text{m})(1.99 \times 10^{-7}\ 1/\text{s})$$

or

$$v_t = 29{,}800\ \text{m/s, or about 30 km/s}$$

That's fast!

PROBLEMS

6.1 A merry-go-round is rotating once every 5 s. How many degrees does it rotate through in half a revolution? How many radians? After 20 rev, how many degrees has it rotated through? How many radians?

6.2 Before there were CDs, there were vinyl records that rotated on a turntable at $33^1/_3$ rev/min. What was their angular speed in revolutions per minute? In radians per second? Vinyl records called *singles* (also called 45s) rotated at 45 rev/min. What was their rotation rate in radians per second?

6.3 The merry-go-round in Problem 6.1 started at rest. If it took 5 s to get it to a rotation rate of once every 5 s, what angular acceleration did the merry-go-round undergo?

6.4 A student tries to open a door at a distance 1 cm from the hinge, at a distance of 10 cm from the hinge, and at a distance 30 cm from the hinge. Why is a larger force required at a distance of 1 cm? How much more force is required at 1 cm than if the exertion point is 30 cm from the hinge?

6.5 Two children are on a merry-go-round at the edge of the platform. They decide to meet in the middle of the merry-go-round. As they walk toward each other in the center, the merry-go-round starts to spin faster, and they are unable to reach the center. Describe what is happening.

6.6 One of the children in Problem 6.5 is located at a distance of 1 m from the center of the merry-go-round, spinning at a rate of once every 5 min. (a) What is the angular velocity of the merry-go-round? (b) What is the tangential velocity of the child? (c) What would be his tangential velocity if he were at the edge of the merry-go-round (at a radius of 3 m)? (d) What would be his angular velocity at the edge of the merry-go-round?

6.7 The planets in the solar system orbit the Sun in elliptical paths. If Venus and Earth have similar masses, and Venus is located at a distance that is approximately three-fourths of Earth's distance from the Sun, what is the ratio of the gravitational force between Venus and the Sun, and Earth and the Sun. That is, what is the ratio of F_G (Venus–Sun) to F_G (Earth–Sun)?

CHAPTER 7

SOLIDS AND LIQUIDS

KEY TERMS

atom, molecule, element, compound, nucleus, electron, solids, alloy, pressure, buoyancy, surface tension, Archimedes' principle

PHYSICS IN THE REAL WORLD

Your body is composed mostly of oxygen (~65% by weight), carbon (~18%), and hydrogen (~10%), three elements that are abundant on the surface of Earth. You are made of tiny building blocks that have been reused an enormous number of times. The hundreds of billions of atoms that constitute you have been a part of Earth for billions of years. The atoms were undoubtedly a part of other plants and animals that walked Earth eons ago. The saying is, "Ashes to ashes and dust to dust," but for all practical purposes atoms are eternal. The atoms that make up you will be part of Earth long after you are gone.

The hydrogen atoms in your body (there are two in every water molecule, and your body is about 75 percent water), have been around since the Big Bang. The carbon, oxygen, and sodium in your cells were made in the core of a star that had as fuel only hydrogen and a little helium, and the atoms of nickel, copper, zinc, and selenium that are rare but essential to the functioning of your cells were generated in the cataclysmic explosion of a star much larger than the Sun. We are physical beings, and the particles that make up our bodies connect us to the universe.

All matter is composed of atoms. Our bodies, the cars we drive in, the houses we come home to—all these objects are composed of atoms. The idea that all things must be composed of fundamental building blocks is a very old one. Greek philosophers proposed the idea more than two thousand years ago. The concept of a smallest, indivisible particle called an *atom* may have been proposed long ago, but it took until the midnineteenth century for scientists to uncover evidence of the existence of atoms and that there are a limited number of "flavors" of atoms (called elements) each of which has unique chemical and physical properties.

ATOMS, MOLECULES, ELEMENTS, AND COMPOUNDS

All things in the universe, from the largest galaxies to the tiniest gnats, are composed of a surprisingly small number of types of atoms. By analogy, we know that the English alphabet is made up of only twenty-six letters; but think of the incredible variety that has arisen from that small number of letters. The richness of language originates from letters (symbolizing sounds) that make words, and words that combine into poems, novels, songs, or political treatises. Atoms are like the alphabet of the universe. All the richness of matter that we observe in our daily lives, and even in the universe out to its farthest reaches, appears to be composed of 90 or so elements; and the overwhelming majority of the visible universe (by mass) is composed of only two elements, hydrogen and helium.

Hydrogen alone accounts for about 90 percent of material that we can observe in the universe, and hydrogen and helium together make up about 99 percent. The amazing thing about hydrogen is that the hydrogen that we observe in the universe has been there since the beginning of time. The universe's simplest atom is also the oldest. Almost all the other elements we know of in the universe were generated in the cores of stars as they cycled through their lifetimes.

The Structure of the Atom

As it turns out, atoms consist mostly of empty space. The tiny *nucleus* of an atom, containing almost all its mass, is but a speck in its total volume. The majority of the volume of an atom consists of a swarm of negatively charged *electrons* moving around the positively charged nucleus in a relatively large volume of space. This volume is sometimes referred to as an *electron cloud.*

The nucleus of the most abundant atom (hydrogen) contains simply a proton, with a mass of 1.67×10^{-27} kg. The nuclei of other atoms contain both protons and neutrons, which are positively charged and neutral, respectively. Neutrons and protons are in turn composed of fundamental particles called *quarks.* We will return to the subject of fundamental particles later in this book.

Moving very rapidly around the nucleus are one or more electrons, negatively charged particles that have about 1/2000 the mass of a neutron or proton. In a neutral atom, the numbers of protons (positive) and electrons (negative) are equal. The ways in which atoms combine with one another is governed mostly by the configuration of these electrons. Atoms can donate, receive, or share their electrons to form molecules.

PHYSICS IN THE REAL WORLD

What is holding you in your seat right now? What holds you up? The atoms in the surface cells of your body are—cushioned by a thin layer of fabric in your clothing—in contact with the atoms in the chair. Your body pushes down on the chair, and the chair pushes up on your body. If the forces are in balance, then the sum of the forces is zero, and you do not move. You are in a static situation.

But what is doing the pushing? Your body is not a block of material, it is composed of atoms—mostly hydrogen, carbon, and oxygen—contained in a dizzying array of molecules; and each of those atoms is made mostly of empty space. Each atom in your body is surrounded by a swarm of electrons (unless the atom is hydrogen, with only one electron), and these electrons, and their negative charge, are what keep you from sinking through your chair. Electrons repel electrons. Because you are mostly empty space, and the chair is mostly empty space, you should sink through it like a ghost. But the electrons that surround all atoms, and bind atoms together as molecules in many cases, push against one another and keep matter "solid."

Molecules

Atoms combine together to form *molecules* as the result of chemical reactions. Burning (oxidation), for example, is a chemical reaction in which atoms or molecules (containing carbon and hydrogen) combine rapidly with oxygen (found in Earth's atmosphere) to generate carbon dioxide and water. Most types of atoms form molecules easily. One family of atoms (called the noble gases) tends not to combine and includes helium, neon, argon, and krypton.

Molecules fill our daily lives. The gas that we inhale consists of diatomic nitrogen (N_2) and oxygen (O_2). A gas that we exhale with every breath contains carbon dioxide (CO_2). Carbon dioxide can be broken into its constituent elements, carbon and oxygen. The combination of these two atoms to create carbon dioxide releases energy, so in nature, the combination is likely to happen.

Types of Bonds

Atoms join together in a number of ways, through the exchange or sharing of the electrons that form their outermost layer. Two of the most prevalent types of bonds are ionic and covalent bonds.

In an *ionic bond,* electrons from one atom are lost to another, and each of the atoms ends up with a net charge. The resulting charge attracts the atoms to one another. An example of this type of bond is found between sodium and chlorine in sodium chloride (salt). Sodium loses an electron (becoming a positive ion) and chlorine gains one (becoming a negative ion), and the two ions are held together tightly by the attraction between the opposite charges.

In a *covalent bond,* one or more electrons are shared between two atoms. Carbon forms covalent bonds with itself and with other atoms, forming everything from diamonds and graphite (carbon bonded to carbon) to carbon monoxide (an exhaust gas from automobile engines).

Elements, Compounds, and the Periodic Table

In the nineteenth century, chemists carefully began to explore the unique characteristics of natural substances, discovering that certain substances—compounds—could be broken down into other substances—elements; and these basic elements were found to have unique physical characteristics.

Elements are materials composed of the same atom and therefore have the same specific chemical and physical properties. If a substance has been broken down by chemical means into its constituent elements, it cannot be broken down any further.

Chemists in the nineteenth century also determined that certain elements have similar characteristics. All the noble gases, for example, were found to be inert, or chemically inactive. Chemists also discovered that elements seem to combine with one another in similar ways. For example, hydrogen combines with both oxygen (O) and sulfur (S) in the same ratios, to form water (H_2O) and hydrogen sulfide (H_2S), respectively. These similarities among elements were first codified in the periodic table (see Figure 7.1).

Not all elements with similar properties occupy the same column in the periodic table (as the noble gases do). For example, metals are grouped together on the left side of the table in more than one column. All metals have similar characteristics: they are good conductors of heat and electricity and tend to have shiny surfaces. Metals tend to form covalent bonds in which the electrons are shared by many atoms, and their atomic nuclei form a lattice structure through which the electrons are free to move. In fact, it is the free movement of electrons in metals that makes metals such good conductors of heat and electricity.

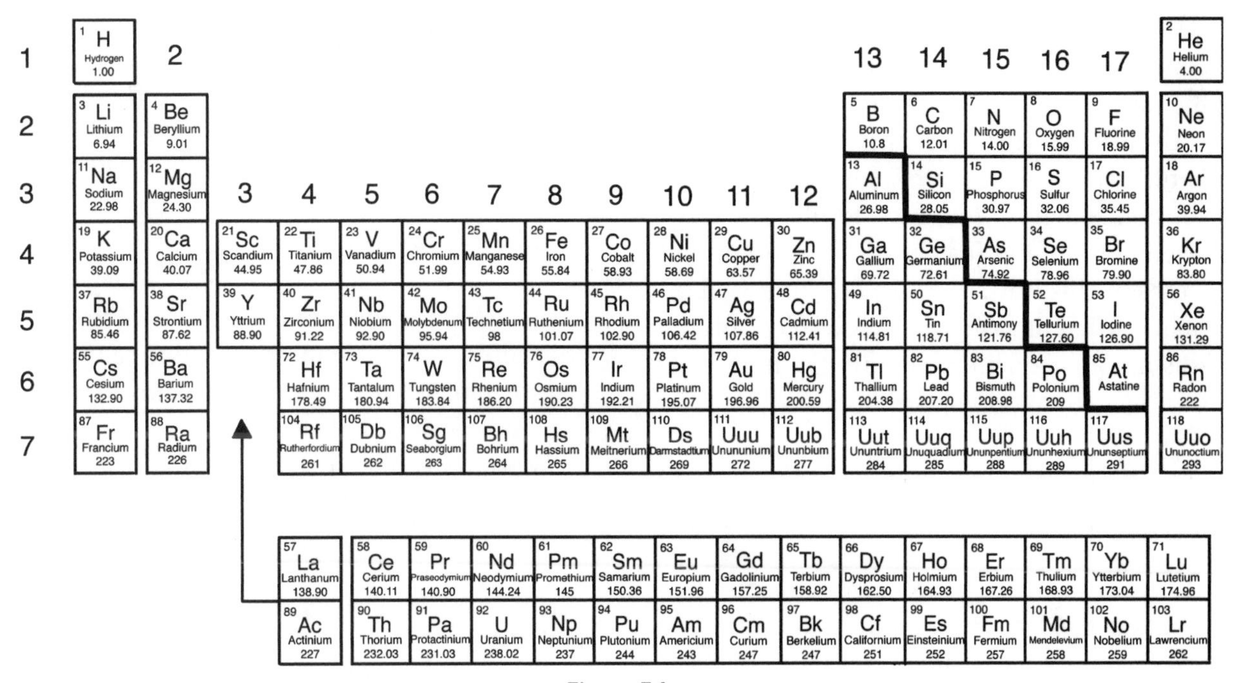

Figure. 7.1
The Periodic Table of Elements

Compounds are substances in which two or more elements are chemically combined. Like elements, compounds have unique characteristics. Water, for example, has a characteristic temperature at which it boils and freezes (at a given pressure) and also has a characteristic density. The water molecule has a somewhat unusual life-giving attribute as well, which is that the solid form of the water molecule (water ice) has a lower density than the liquid form (see Table 7.1). As a result, when bodies of water freeze, they "freeze over" from the top down, effectively insulating the liquid water below. If water ice were denser than liquid water (as is the case for most compounds), bodies of water would freeze solid from the bottom up, dooming any life in it.

It was a great stride forward in understanding to determine that the same element can be a component of molecules (and therefore compounds) having very different properties. Hydrogen, for example, is found in water

Substance	Density (g/cm^3)	Density (kg/m^3)
wood, pine	0.5	500
gasoline	0.70	700
ice	0.92	920
water	1.00	1,000
seawater	1.03	1,030
aluminum	2.7	2,700
limestone	3.2	3,200
iron	7.90	7,900
brass	8.6	8,600
copper	8.9	8,900
lead	11.3	11,300
mercury	13.6	13,600
gold	19.3	19,300

Table 7.1 Densities of Some Common Materials

(H_2O), but also in ammonia (NH_3). When combined with oxygen, hydrogen generates the basis of life on Earth, but when combined with nitrogen (the most abundant element in Earth's atmosphere), it creates a very different substance that is often toxic to life. The four "elements" of old (earth, air, fire, and water) were conclusively shown to be composed of many different atoms and molecules, and fire was not an element at all, but the manifestation of a chemical reaction!

SOLIDS

Materials in which the relative location of atoms is fixed, owing to bonds and attractive forces among them, are called *solids*. Many elements, molecules, and compounds are solids at room temperature on the surface of Earth. Iron, for example, is a solid, as are carbon and lithium. Table salt (sodium chloride) is a compound that is a solid at room temperature; in fact, it is what is called a crystalline solid, meaning that the sodium and chlorine nuclei arrange themselves into a fixed structure. The water molecule is a compound that is sometimes a solid, in the form of water ice, at the surface of Earth, depending on location and season.

In general, solids can be characterized as either *amorphous* or *crystalline*. The arrangement of atoms in an amorphous solid is random, whereas the arrangement in a crystalline solid is highly ordered. The atoms in glass (a compound containing silicon) are randomly arranged and can even shift very slowly, and the atoms in a diamond are highly ordered. That water ice is slightly less dense than liquid water is the result of the randomness of the molecular arrangement in a liquid, and the more ordered arrangement of water molecules in water ice.

Density

The spacing between atoms in a substance determines its density. In general, we know that density is equal to mass divided by volume, or

Density = Mass/Volume

As a result, the units of density are kilograms per cubic meter, or grams per cubic centimeter (see Chapter 2 for a more complete description of density). The density of an element or compound is one of its identifying physical characteristics. Table 7.1 gives densities for some common substances.

Alloys

An *alloy* is a solid solution of one or more metals, or a metal and a nonmetal. Bronze is an alloy of copper and tin, for example. Steel is an alloy of iron and carbon. In ancient times, coinage was made from elements (copper, silver, gold). Today, coins are made from more durable (and less expensive) alloys. Alloys are formed by mixing the component materials in a molten state, and allowing the mixture to cool. Many of the materials used in modern-day life are alloys, including many of the metals in cars and airplanes, and the entire class of materials called semiconductors.

LIQUIDS

In a liquid state, atoms are not in fixed positions relative to one another but can move freely. The surface of Earth is covered in a magnificent ocean of liquid water, and water (or some other liquid) may be essential for the chemical mixing that must take place for life to arise.

Pressure

Liquids are composed of atoms or molecules that are in constant motion, and these particles exert forces on their containers. For many applications involving liquids, it is convenient to quantify the ratio of the force exerted per unit area as *pressure;* that is,

Pressure = Force/Area

Because force is measured in newtons (N) and area in square meters, the unit for pressure is newtons per square meter, or pascal (Pa). One can exert large pressures by reducing the amount of area over which a force is exerted. High-heeled shoes, for example, exert the force of a woman's weight over a small area, resulting in higher pressures. Snowshoes have the opposite effect, spreading the weight of a person out over a larger area, resulting in a lower pressure.

In a liquid, pressure increases with depth. In fact, the pressure exerted by a liquid at a given depth depends on two quantities: the density of the liquid and the depth in the liquid. Submarines and other submersible ships have to be able to withstand very high pressures, for the deeper they go, the more external pressure pushes on them.

The pressure at some depth in a liquid of a given density is given by:

$$\text{Pressure} = D \times g \times h$$

where D is the density of the fluid, g is the acceleration of gravity, and h is the depth in the liquid. For example, at a depth of 10 m, the pressure in the ocean is given by

$$\text{Pressure} = (1000\ \text{kg/m}^3) \times 9.8\ \text{m/s}^2 \times 10\ \text{m}$$

$$\text{Pressure} = 9.8 \times 10^4\ \text{N/m}^2$$

One atmosphere is the equivalent of 1.013×10^5 N/m^2, so this means that the water above an object at 10 m adds almost an additional atmosphere of pressure. This pressure increases linearly with depth, so at 100 m (10 times deeper), there are 10 additional atmospheres of pressure.

Buoyancy

If you have ever been in a swimming pool or in the ocean, you may have noticed another effect of liquids, namely, a force referred to as the *buoyant force.*

> **EXPLORATION 7.1**
>
> Take two Styrofoam cups and fill them both to the top with water. Given what you now know about pressure, if you poke a hole in each cup, one near the top of one cup, and one near the bottom of the other cup, which stream will be under greater pressure and thus leave the cup at higher velocity? Try it out and see.
>
> It turns out that the water at the bottom of the cup is under greater pressure (exerting a greater force per unit area), and so it leaves the cup with higher velocity.

Buoyancy is very much related to the pressure increase with depth. Because pressure does increase with depth, the downward pressure on the top of an object is less than the upward pressure on the bottom (see Figure 7.2). Thus, there is a net upward force exerted on any submerged object in a liquid.

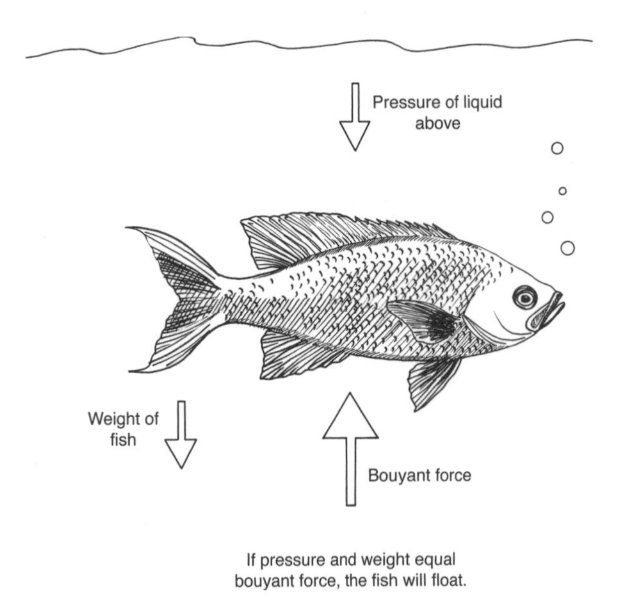

Figure 7.2

When an object is placed in water, it will float in the water, float completely submerged, or sink to the bottom. The final location of the object depends on the balance between its weight and the buoyant force exerted on it.

The buoyant force exerted on an object turns out to be equal to the weight of the volume of liquid displaced. This statement is sometimes called *Archimedes' principle,* after the ancient Greek scientist Archimedes (287–212 B.C.E.), who discovered this relationship. In the case of a rock, a rock thrown into a pond is buoyed up by a force equal to the volume of the water it displaces. Because a rock is far denser than water, even when fully submerged, it does not displace a volume of water equal to the weight of the rock. Thus, although a rock will weigh less under water (as you may have noticed if you have ever tried to move large rocks around underwater), it will not float.

Styrofoam, however, is less dense than water, so it will float even when a small part of it is submerged. It has to displace only a small amount of water in order to displace a volume equal to its weight. This is why the surface of Earth's oceans is littered with chunks of Styrofoam and not pebbles.

A brief reminder about mass and weight: the mass of an object does not change in any of

PHYSICS IN THE REAL WORLD

When you get into a swimming pool, you feel lighter. Why is that? The reason for your apparent lightness is that the water around you is exerting an upward force on you that partially balances the downward force of gravity. This effect is the reason why astronauts train in deep pools of water. They can be sufficiently loaded with material to achieve neutral buoyancy, meaning that they feel as if they are weightless in the watery environment. NASA finally settled on this as the best method for training astronauts to work in the weightless environment of Earth orbit.

EXPLORATION 7.2

Can an egg float in water? It depends on how salty the water is. Take a glass of water and slowly lower a fresh, uncooked egg into it. The egg should sink to the bottom of the glass. Now slowly add salt to the water, dissolving it by mixing. You may have to carefully remove the egg to dissolve the salt completely. Once you have added a sufficient amount of salt, the egg will become neutrally buoyant and will eventually float. By adding salt to the water, you have increased its density. This increase in density accounts for the increased buoyancy of saltwater versus freshwater.

EXPLORATION 7.3

How does the density of a duck compare with the density of water? How do you know? How does the density of a fish compare with the density of water? How do you know?

Any object that naturally floats on the surface of water must be less dense than water, so the density of a duck is less than that of water. A fish maintains *neutral buoyancy* most of the time, so it must have a density close to that of water.

these circumstances. The weight of an object can change owing to the buoyant force exerted by water, but the mass of the object (you, a duck, a fish) always remains the same.

How, you might ask, does something denser than water ever float? Ocean liners, as we all know, are not made of Styrofoam but of far more dense steel. The secret is that the steel is formed into a shape that displaces a large volume of water. Thus, if a ship weighs 50,000 metric tons, it must displace 50,000 metric tons of water to float. Boats of all types (canoes, rowboats, oil tankers) are designed to displace the necessary amount of water.

Surface Tension

You may have at some point in your life watched a bug suspend itself on the surface of a liquid. Water striders (also called *pond skaters*) skate across the surface of creeks. Ants that fall into the sink are able to hover on the surface, apparently borne up by some force that doesn't seem to help us when we jump into the pool. How can this be so?

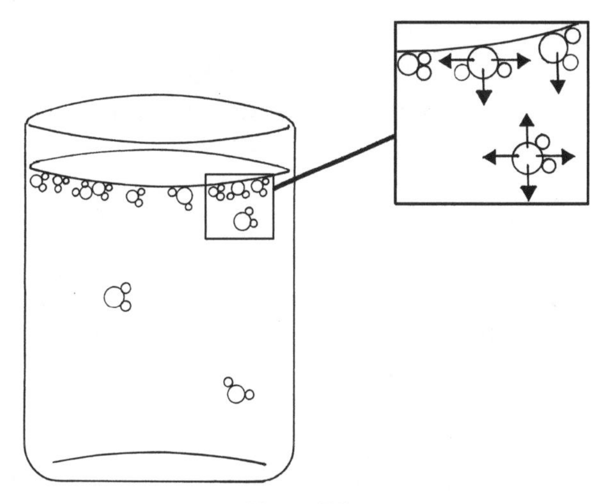

Figure. 7.3

Imagine a water molecule at the surface of a glass of water (see Figure 7.3). Water molecules are *polar,* meaning that they have a slightly positive charge on one side (the hydrogen atom side) and a slightly negative charge on the other (the oxygen atom side). Thus, all water molecules exert attractive forces on one another and on other polar molecules. Water molecules located centrally in the glass are pulled in all directions by neighboring molecules, but molecules at the surface are pulled only downward. This downward attractive force stretches out and minimizes the surface area of the water.

The same effect is responsible for the way liquid droplets form into spheres. If a droplet is sufficiently small, the minimizing effect of surface tension creates a tiny sphere of water.

PROBLEMS

7.1 Atoms are composed of neutrons, protons, and electrons. Which of these three is the most massive? What is the charge on each of these components of an atom?

7.2 Describe the difference between an atom and a molecule; between an element and a compound. What is the origin of the elements in our bodies?

7.3 How many atoms are there in the body of a person with a mass of 75 kg. For simplicity, assume that the body consists of hydrogen atoms, each of which has a mass of 1.67×10^{-27} kg.

7.4 A person has a mass of 75 kg and a volume of 0.064 m^3. Will she sink or float in water? (*Hint:* The density of water is 1000 kg/m^3).

7.5 Would a lead ball sink or float in a pool of mercury? Would a gold ball sink or float in a pool of mercury? How do you know?

7.6 Devise an experiment to determine whether a penny is made of pure copper or an alloy.

7.7 Does pressure increase or decrease as you move higher into Earth's atmosphere? Why is this?

7.8 If the person in Problem 7.4 jumps into a pool of fresh water, what upward buoyant force will she experience? What buoyant force would she experience if, hypothetically, she were to jump into a pool of mercury?

HEAT, TEMPERATURE, AND STATES OF MATTER

KEY TERMS

temperature, absolute zero, heat, calorie, specific heat, latent heat, coefficient of linear expansion, laws of thermodynamics, ideal-gas law, conduction, insulators, convection, radiation

PHYSICS IN THE REAL WORLD

Sitting alone in your kitchen on a hot summer afternoon, you place some ice cubes in a glass of water. They float (being less dense than the water) but with most of their volume under the surface of the water.

The ice cubes are colder than the water. The water is warmer than the ice cubes. As you sit there watching the glass of water the ice cubes slowly melt as the jiggling water molecules in the glass collide with the water molecules in the ice cubes, breaking up the crystalline structure of the water ice. As energy is transferred from the water (warmer) to the ice (cooler), and assuming that the room you are seated in is above the temperature at which water freezes, the two materials will come to an equilibrium temperature, and the ice cubes will eventually melt completely. As you sit there sweating you recall that since the floating ice cubes are displacing a volume of water equal to their weight, the melting of the ice cubes will not cause the level of the water to rise at all.

All matter is composed of atoms, and all atoms are in constant motion. Atoms in a liquid are clearly moving past one another when fluids flow, but even in a solid material, atoms are in motion. The atoms that form the lattice structure of a crystalline solid like table salt are in motion, and we measure the energy of this motion as *temperature*. When two bodies in thermal contact (like ice cubes in a glass of water) have different temperatures, thermal energy can be transferred between them.

TEMPERATURE AND ITS MEASUREMENT

We are all familiar with temperature as a measure of how hot or cold it is outside or inside our home. When we measure temperature, we are measuring the mean hotness or coldness of the air outside, or of the air in our home. Fundamentally, temperature is a measure of how much atoms or molecules are jiggling around, how much random motion they have. Atoms and molecules that have a lot of random motion are hotter than those that have less, and when the random motion of a given element or compound gets sufficiently small (that is, the material is sufficiently cold), then it may be able to solidify or "freeze" into an amorphous or crystalline solid.

The nerve endings in our bodies are sensitive to variations in temperature, and even without a thermometer, your body will tell you whether you are warm or cold; but temperature can be accurately measured with a thermometer. A thermometer is often a narrow channel in a tube of glass that is inscribed with a calibrated scale, and the tube is filled with a thin column of mercury. When the mercury is warmer the atoms have more random motion and the mercury expands, registering a higher temperature. When the mercury is cooler its atoms have less random motion, the mercury contracts, and the thermometer records a lower temperature.

Celsius, Fahrenheit, and Kelvin

There are three scales in which temperature is generally measured, the *Celsius* (C) scale, the *Fahrenheit* (F) scale, and the *Kelvin* (K) scale. In the United States, temperatures are generally reported in degrees Fahrenheit (°F), whereas in the rest of the world they are reported in degrees Celsius (°C). As with units of length and weight, the United States has largely ignored the metric system used by the rest of the world.

On the Celsius scale, used in scientific work, at a pressure of 1 atmosphere (the average pressure found at the surface of the Earth), the temperature of a mixture of water and ice (the freezing point) is designated as 0 degrees (0°C), and the temperature of water that is boiling to produce steam (the boiling point) is 100 degrees (100°C). The temperature range between these two marks is divided into 100 equal parts, each being equal to 1 degree Celsius (1°C).

The system more familiar to U.S. readers is the Fahrenheit scale. In this system, the temperature of the freezing point of water is 32 degrees (32°F), and the temperature of the boiling point of water is 212 degrees (212°F). One can easily convert between these two temperature scales using the formula

$$(\text{degrees})\text{F} = \frac{9}{5}(\text{degrees})\text{C} + 32$$

EXPLORATION 8.1

If the temperature on either temperature scale is below zero, remember to place a minus sign in front of that temperature in the equation. For example, the temperature at which carbon dioxide freezes (also known as *dry ice*) is about −80°C. How would you find the equivalent temperature in degrees Fahrenheit?

Use the conversion formula given, inserting the temperature as −80°C

$$°\text{F} = \frac{9}{5}(-80)\text{C} + 32 = -144 + 32 = -122°\text{F}$$

Thus the freezing temperature of carbon dioxide is −122°F. That frozen carbon dioxide has been found at the north and south poles of Mars should tell you something about the surface temperatures of that planet.

The other temperature scale most commonly used in science (both in physics and astrophysics) is called the *Kelvin* or *absolute scale*. This scale is based on the existence of *absolute zero*—the temperature at which, in theory, atomic motion would stop. The zero point on the Kelvin scale, abbreviated 0 K, is set at absolute zero. The graduations on the Kelvin scale are the same size as a degree on the Celsius scale; only the zero point is different.

As we mentioned previously, temperature is the result of random motions in the atoms that make up a material. If this random atomic jiggling stopped entirely, the material would be at 0 K. At this temperature, a gas would also exert no pressure.

Absolute zero is the lowest temperature possible in the universe—in principle, there is no upper limit to temperature. As a result, on the Kelvin scale there are no negative temperatures. The freezing point of water on the Kelvin scale is 273 K, and the boiling point is 373 K. By international agreement, the increments on the Kelvin temperature scale are simply called *kelvin*; a reading in kelvin can be converted to degrees Celsius simply by adding 273. For example, a summer day's temperature of 27°C would be 300 kelvin (27 + 273), or 300 K.

HEAT AND SPECIFIC HEAT

Heat is thermal energy that can be transferred between two bodies that are at different temperatures. Heat naturally flows from a body at a higher temperature to a body at a lower temperature. The amount of thermal energy that a body contains depends on two quantities, its temperature and the amount of material it contains, that is, its mass. A tiny speck of molten metal may have less thermal energy than a bucket of hot water. And the molten speck, as a result, will cool off much more quickly.

The common unit used for heat is the *calorie.* A calorie is defined as the amount of heat required to change the temperature of 1 g of water by 1°C. A *kilocalorie* (C) is 1000 calories, and the ratings of calories in foods are actually in kilocalories. Calories are a unit of energy, and the energy content of food could just as easily be measured in joules (J).

Other substances generally require less h eat than does water to change their temperature by 1°C. We know this from experience. It takes a bowl of soup—made mostly of water—a lot longer to cool off (come into equilibrium temperature with its surroundings) than it does for a piece of toast. We say that soup (water) has a higher specific heat than toast.

The *specific heat* of a substance is the amount of heat required to raise the temperature of 1 g of the substance by 1°C. Water absorbs more heat than most other substances before its temperature increases by 1° and, as a result, can burn your tongue for a while after it has been heated! This property of water serves to moderate temperature changes in regions near a large body of water such as a lake or an ocean. Conversely, desert environments experience much more extreme temperature changes, due to the lack of temperature-moderating water vapor.

The amount of heat required to raise the temperature of a substance can be calculated if you know the specific heat of the substance. In general, the amount of heat required to change the temperature of a substance is proportional to the mass of the substance and the change in temperature, according to the following relation:

$$Q = mc\,\Delta T$$

where Q is the heat required (in joules or calories), m is the mass of the substance, c is the specific heat of the substance, and ΔT is the change in temperature. Table 8.1 gives the specific heats for a variety of substances at 20°C and 1 atm pressure.

Substance	Specific Heat (J/kg•°C)
air	1050
ethyl alcohol	2430
aluminum	920
glass	840
soil	1050
wood	1680
water	4186

Table 8.1 Specific Heats of Various Substances

PHASE CHANGES, OR CHANGES OF STATE

As we have mentioned, the phases of matter are solid, liquid, and gas. The phase of a material depends on two factors: its internal energy and the external pressure.

Surprisingly, when substances change phase (from solid to liquid, for example) they absorb or give off energy in the form of heat, yet there is no temperature change. Thus, a certain amount of heat must be added to a sample of ice of a given mass before it will all melt, but as it melts, its temperature will stay at 0°C. Therefore, the preceding equation for temperature changes does not apply.

The heat involved in a change of phase is called *latent heat* and is the amount of heat (energy) required to change the phase of 1 kg of a given substance.

For example, the latent heat of water at the freezing point (called the *latent heat of fusion*) is 3.33×10^5 J/kg, and the latent heat of water at the boiling point (called the *latent heat of vaporization*) is 22.6×10^5 J/kg. Notice that far more energy (about seven times) is required to separate water molecules from typical distances in a liquid to the distances in a gas than is required to break the lattice structure that distinguishes ice from water. This enormous amount of energy is the reason why steam burns are so dangerous and severe. When steam condenses, a large amount of energy is deposited on the skin (if that is where it is condensing).

PHYSICS IN THE REAL WORLD

One way to think about the factors of temperature and pressure involves the planet Mars. There are places on the surface of Mars where it is warm enough for liquid water to exist (that is, the temperature is "above freezing" as we would refer to it on Earth); however, the pressure of the atmosphere on the surface of Mars is very low, about 1/100 what it is on the surface of Earth. Therefore, liquid water cannot exist on the surface of Mars, not because it is not warm enough—certain places are—but because the pressure is too low to "hold" water in its liquid state.

THE BEHAVIOR OF SOLIDS AND LIQUIDS

When a solid or a liquid substance is heated its temperature increases. The result is that the atoms that compose it move around more rapidly, colliding with one another more frequently, pushing molecules farther apart on average, and the volume of the substance generally increases. The cracks apparent in freeway asphalt (especially in northern climates) are evidence that solid materials do indeed expand and contract with temperature changes.

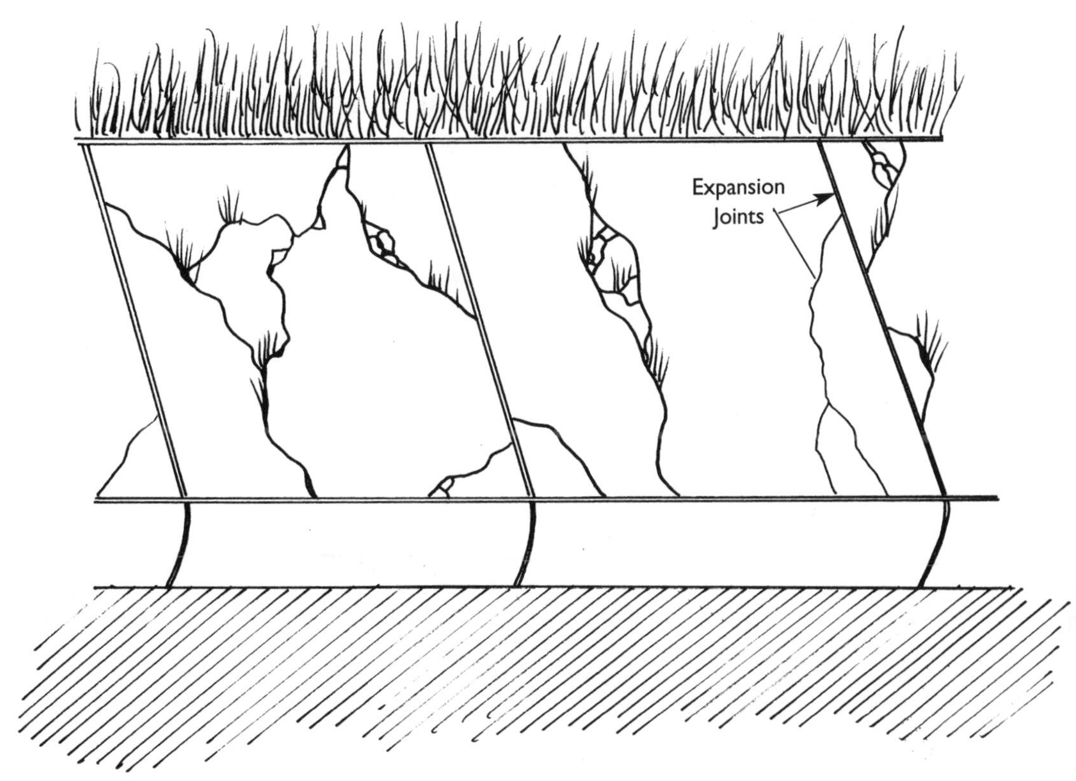

Figure 8.1
Sidewalks can crack due to repeated expansion and contraction.

Coefficient of Linear Expansion

Heating a substance involves increasing the average energy of its molecules; cooling a substance removes this energy. Because increasing the average internal energy of molecules increases their average separation, substances generally expand when they are warmed, and contract when they are cooled, but they do so at different rates.

Solid objects increase in length by a certain fraction for each rise in temperature. This fraction, which can be converted to either degrees Fahrenheit or degrees Celsius, is called the *coefficient of linear expansion*.

The change in length of a given material is given by the formula

$$\Delta L/L = \alpha \Delta T$$

where ΔL is the change in length, L is the original length, α is the coefficient of linear expansion, and ΔT is the change in temperature. The coefficients for a small number of substances are given in Table 8.2.

Because the change in length is proportional to the original length, changes observed for

Material	Coefficient (1/°C)
aluminum	0.000024
brass	0.000019
iron or steel	0.000012
ordinary glass	0.000009
Pyrex™ glass	0.0000033

Table 8.2 Coefficients of Linear Expansion for Selected Materials

small lengths of a material are rather small, but over long distances, the size of a bridge for example, the changes in length can be significant. For example, a 1-m length of steel whose temperature rises from the freezing point of water (0 °C) to 40 °C will expand by

$$\Delta L = L\alpha\Delta T$$

$$\Delta L = 1 \text{ m} \times (0.000012\ 1/°\text{C} \times 40°\text{C})$$

$$\Delta L = 4.8 \times 10^{-4} \text{ m, or } 0.048 \text{ cm}$$

which is a very small change in length—less than 1 mm. The change in length of a 1000-ft steel bridge, however, will be a significant factor in its design.

EXPLORATION 8.2

Suppose you were designing a 1000-ft-long steel bridge. How much expansion would you need to allow for between an ambient winter temperature of −10°C and a summer temperature of +40°C?

The temperature change (ΔT) you must consider is 50°C. Using an expansion coefficient for steel of 0.000012/°C, the total expansion (and contraction) would be 0.000012 × 50 × 1000 ft, and you find that your bridge will expand and contract between the extremes of temperature by a total of 0.55 ft. Clearly, expansions at this level must be engineered into the design in what are called *expansion joints*. (See Figure 8.1 on previous page.)

Most liquids expand and contract with temperature the same way that solids do, but water is an exception. Between the freezing point and about +4°C, water *contracts* very slightly. At temperatures higher than +4°C, water again expands like a normal liquid. This property of water, together with the fact that it freezes at a moderate temperature, makes water vital to life (see the discussion in Chapter 7).

THE LAWS OF THERMODYNAMICS AND IDEAL GASES

Thermodynamics

Thermodynamics is the study of the relationship between heat and mechanical energy. Thermodynamics is a science that predates a detailed understanding of the details of atomic and molecular structure and is founded on energy conservation principles. There are two fundamental *laws of thermodynamics*, based on empirical observations of physical systems.

The *first law of thermodynamics* states that adding heat to a system or doing work on it results in an increase in the internal energy of a system; conversely, if the system does work or if heat is removed from it, its internal energy decreases. The basic idea here is that energy must be conserved when heat (energy) is added to or removed from a system. In a simple sense, the first law of thermodynamics is a version of the law of conservation of energy.

The *second law of thermodynamics* states that heat flows spontaneously from a hot body to a cold body. It is a statement about the natural direction in which heat (energy) flows. The second law of thermodynamics is also sometimes called the *law of entropy*, since in general, natural systems proceed to states of greater disorder. *Entropy* is a measure of the amount of disorder in a system—flowers whither, dishes break, mountains erode—and the entropy of a system increases with time.

Ideal Gases and Gas Laws

An *ideal gas* is a gas in which the molecules or atoms can be thought of as individual, point-like particles that do not exert intermolecular forces on one another. Of course, no gases are truly ideal, but air in our atmosphere and many gases under normal conditions behave like ideal gases—that is, they expand when heated and contract when cooled, much like solids and liquids do.

For an ideal gas at constant temperature the ratio of volume to its Kelvin temperature remains a constant. This law can be stated as follows:

V/T = constant, or $V_1/T_1 = V_2/T_2$

This law is sometimes referred to as *Charles's law* (after Jacques Charles, 1746–1823). Also, for an ideal gas at constant temperature, the product of pressure and volume is a constant, or

PV = constant

or

$P_1V_1 = P_2V_2$

which is referred to as *Boyle's law* (after Robert Boyle, 1627–1691). The two laws can be combined into what is referred to as the *ideal-gas law*, or

$P_1V_1/T_1 = P_2V_2/T_2$

The ideal-gas law is also sometimes written as

$PV = nRT$

where the pressure (P) and volume (V) of an ideal gas are related to its Kelvin temperature (T), the number of moles of the gas (n) and a constant called the universal gas constant R, (where R = 8.3145 J/mol•K).

NOTE: A mole is the number of particles (atoms or molecules) contained in a sample of element or compound with a mass in grams equal to the atomic or molecular weight. This mass contains what is called Avogadro's number (6.0221367×10^{23}) of atoms or molecules. For example, a mole of carbon atoms has a mass of 12 g and contains 6.022×10^{23} carbon atoms.

EXPLORATION 8.3

A helium balloon has a volume of 0.1 m^3 and a temperature of 27°C. If the temperature of the balloon rises to 42°C when it is left in a car with the windows rolled up, what will be its new volume?

Using Charles's law (and converting temperatures from degrees Celsius to Kelvin), we say that

$V_1/T_1 = V_2/T_2$

$(0.1\ m^3)/300\ K = V_2/(315\ K)$

or $V_2 = 0.105\ m^3$

This is an increase in volume of 5 percent and, depending on the strength of the balloon material, it could cause it to pop!

In a parallel to our description of the buoyancy of objects in water, objects can also be buoyant in air. An object "submerged" in Earth's atmosphere will feel more pressure on its bottom side than on its top side (for the same reasons as described for an object submersed in a liquid), resulting in a net upward force. If the weight of a container (say a hot air balloon) and its contents (hot gas) are less than the weight of the air displaced, then the balloon will float (see Figure 8.2).

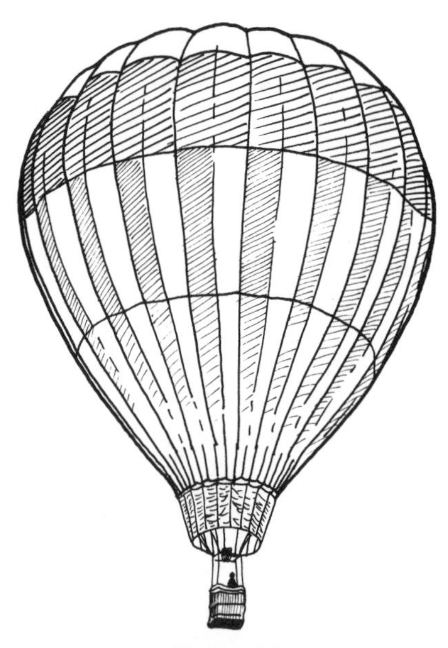

Figure. 8.2
A hot air balloon is "buoyant" in air.

PHYSICS IN THE REAL WORLD

Sometimes a car has a sheet of plastic covering a broken window. When the car is traveling at highway speeds, this sheet is often seen bulging out. It bulges owing to the high velocity of the air outside the car versus the still air inside the car. The higher velocity outside means a lower pressure, and the plastic bulges outward. A similar effect is also seen when a truck travels at high speed with a tarp covering the truck bed.

Unfortunately, a high-velocity wind blowing over the roof of an airtight house can make the roof go airborne if the wind velocity is sufficiently high.

Another aspect of gases that has many applications in everyday life is referred to as *Bernoulli's principle.* An eighteenth-century Swiss scientist, Bernoulli observed that there is a relationship between the velocity of a fluid (or a gas) and its pressure. In particular, he found that the pressure of a fluid

EXPLORATION 8.4

Take a narrow strip of lightweight paper and hold it in front of your mouth, below your lower lip, parallel to the floor. Carefully blow across the top of the paper. The paper should rise up as a result of the higher velocity of the air above the paper versus below it. The higher velocity on the top side of the paper results in a lower pressure and consequently creates an upward force on the paper.

decreases as its velocity increases. In some ways, this principle is again an energy conservation law. There is some energy in the internal pressure of a fluid (related to its pressure), and some energy related to its motion (kinetic energy). If the velocity of a fluid increases (thus increasing its kinetic energy), then its internal energy associated with pressure must drop.

CONDUCTION, CONVECTION, AND RADIATION

Heat can be transferred between substances in a number of ways, including by conduction, convection, and radiation.

Conduction

We previously stated that heat is simply a transfer of thermal energy between two objects, energy associated with the random motion of molecules.

In this picture, when two objects are put in contact, the faster-moving molecules of the warmer object collide with the slower-moving molecules of the colder object, transferring some of their energy to the molecules in the cooler object. The warmer object loses energy (drops in temperature) while the cooler one gains energy (rises in temperature) until the two objects or substances are at the same temperature. This type of heat transfer, in which collisions between molecules transfer heat, is called *conduction*. All materials conduct heat at different rates, and metals are some of the best conductors. Stone is a moderately good conductor, and wood, paper, cloth, and air are relatively poor conductors.

PHYSICS IN THE REAL WORLD

You are about to lift the pot of boiling water from the stove when you notice that the handles are made of metal and are directly attached to the pot. You pause, remove two oven mitts from the drawer, and place one on each hand before proceeding.

You are instinctively making use of different rates of conduction for different materials when you do this. Manufacturers of pots and frying pans make use of the differing conductive properties of materials, too, when they encase metal pot handles in wood or in another material that does not conduct heat well, including some high-density plastics.

Materials that conduct heat poorly are called *insulators*. Air, for example, makes an excellent insulation material when it is trapped between spaces, as in double-pane windows that trap air between two panes of glass, and is used to increase the insulation value of a home. Wool clothes, synthetic foams, and materials filled with loose fibers, feathers, or down (like a comforter) insulate by means of the air trapped within them; and when you naturally layer your clothes on a cold day, it is the air trapped within them that decreases the efficiency of the transfer of energy from your warm body to the cold air.

Made famous by manufacturers like Thermos™, vacuum bottles make use of the perfectly nonconductive properties of a vacuum. Such bottles contain a double-walled inner liner that is pumped clear of air and then sealed. The same principle keeps the contents either hot or cold: thermal energy cannot be transferred through the vacuum barrier.

Convection

Because of their relatively low density and smaller number of molecular and atomic collisions, most liquids and all gases are poor conductors; however, gases and liquids can transfer heat in another way: through *convection*, which is the movement or circulation of parcels of heated liquid or gas.

Consider this example: Directly above a bonfire, the air is warmed, causing it to expand and become less dense than the surrounding air. The air above the fire rises. Cooler air

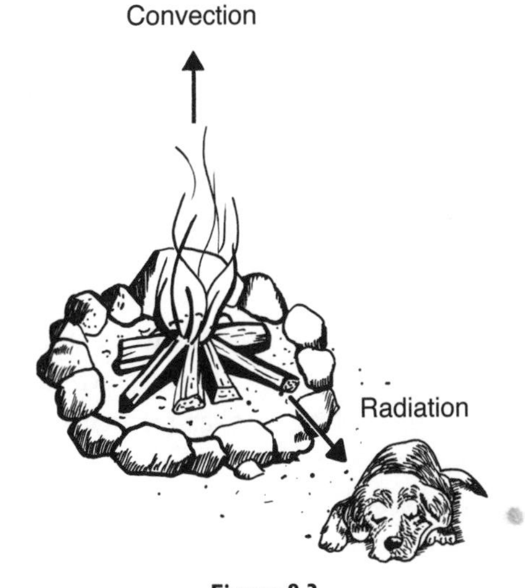

Figure 8.3

flows in from the sides to take its place, and soon a continuous circulation of air is set up. This type of convection can be deadly to firefighters in a forest fire, because it rapidly removes oxygen from the surrounding area.

Convective cycles occur in Earth's atmosphere, giving rise to winds, and also in Earth's oceans, giving rise to currents. Convection is also one of the ways that heat makes its way out of the interior of the Sun to its surface.

Thermal Radiation

If you were sitting next to the bonfire in the preceding convection example, you would feel the warmth of the fire (see Figure 8.3 on previous page). Yet, that heat is not reaching you by convection—we know that the hot air is in fact going straight up and being replaced by cooler air flowing in from the sides.

Also, the heat conductivity of air (a gas) is very low. So how do you feel the warmth of the bonfire? Energy is leaving the fire through the process of *radiation* and is being absorbed by your body. *Radiation* is simply the transfer of energy from one point in space to another through the oscillation of electromagnetic fields. The motion (jiggling) of the electrons in the object that is emitting radiation generates electromagnetic waves. These waves are transmitted through empty space, and when they strike your body, the electrons in your body absorb the radiation; the atoms in your body start to move around at a greater velocity, and you warm up. The electrons act like a pebble tossed into a lake. We may detect the waves that the pebble makes far from where the pebble enters the water. In a similar fashion, we can see the electromagnetic waves caused by jiggling electrons, even though the electrons doing the jiggling are far away. In this way, electromagnetic waves can carry energy between two points that are not connected physically. So, unlike conduction and convection, radiation does not require direct contact between two substances in order to transfer heat.

All objects that have a temperature above absolute zero radiate, and hotter objects radiate far more energy than cooler ones. This phenomenon is most clear with the ultimate bonfire, the Sun. Located 1.5×10^{11} m away from us, the surface of the Sun radiates energy at a variety of wavelengths that peak in the visible part of the spectrum.

PHYSICS IN THE REAL WORLD

Almost all the energy available to us on Earth (aside from geothermal energy) comes to us directly or indirectly from the Sun. When we burn coal or petroleum, we are just releasing potential chemical energy from the Sun that was stored in plants millions of years ago. Even a hydroelectric power plant in which falling water turns the turbines that generate electricity derives its energy from the cycle of evaporation and condensation maintained by the Sun: water is lifted from lakes and oceans and condensed into rain that feeds streams and waterfalls.

The Sun is an extremely hot body—its surface temperature is around 6000 K—and so it radiates enormous amounts of heat; however, all objects—buildings, trees, and your body—radiate heat into their surroundings. The air in a crowded theater can become warm because each person in it is radiating heat equal to that given out by a 75-W lightbulb; night-vision goggles, which are sensitive to

PHYSICS IN THE REAL WORLD

The greenhouse effect is a result of the absorption and partial reradiation of solar energy at the surface of Earth. This effect is in fact required for life on Earth. Without the greenhouse effect the surface temperature of Earth–due to its distance from the Sun–would be so cold that all water could freeze; however, the surface of Earth (and some gases in its atmosphere, including carbon dioxide and water) causes Earth to retain some of the radiation that it absorbs. As a result Earth warms up slightly above the temperature it would have were there no atmosphere. The word greenhouse is used for the following reason: The rays of the Sun pass readily through the glass of a greenhouse, warming the soil inside. The soil, being warmed, also emits radiation, but at a longer wavelength. These longer wavelengths cannot get back out of the greenhouse through the glass, and the greenhouse acts as a heat trap.

Earth's atmosphere (mostly molecular nitrogen and oxygen, with a tiny fraction of CO_2 and H_2O) works like the glass in the greenhouse. Visible radiation from the sun passes through the atmosphere to warm Earth. Earth re–reradiates longer wavelength (infrared) radiation that cannot completely escape the atmosphere, particularly because of the carbon dioxide in it, therefore, some of the heat is returned to Earth's surface. The higher the amount of carbon dioxide in the atmosphere, the more heat is retained. The burning of fossil fuels can raise the carbon dioxide level of the atmosphere, thus trapping even more heat.

infrared radiation, allow you to see the heat radiation given off by people around you.

PROBLEMS

8.1 It is a summer day in Atlanta and the temperature is 100°F. What is the temperature in degrees Celsius? In Kelvin?

8.2 A scientist measures the temperature of a substance to be 100 K, if the temperature rises by 2 K, how many degrees Celsius has it gone up? How many degrees Fahrenheit?

8.3 Why is hot water a good choice for filling a bottle to keep your bed warm at night? Why would it not be a good choice to put that hot water in an insulated thermos bottle if you wanted to feel the warmth?

8.4 Which requires more energy, raising the temperature of a 1-kg block of aluminum by 20°C, or raising the temperature of 500 g of water by 5°C?

8.5 How much energy is required to melt a 1-kg block of ice?

8.6 The steel panel of a car door is 1.00 m long. Assuming that it expands in this direction as described in the text, what is a safe separation between the door panel and the body of the car that will allow the panel to expand?

8.7 How does one explain the ordered nature of living things if the second law of thermodynamics is correct?

8.8 If you have a helium balloon in the car and you suddenly accelerate from a stop, you might expect the balloon to go to the back of the car. If you do the experiment, however, you will find that it moves to the front of the car. Can you explain this phenomenon using what you know about ideal gases?

8.9 In the *Physics in the Real World* example about a house losing its roof in a condition of high winds, why is it stated that the house is airtight? Would it make you safer to open the windows?

8.10 Give an example of heat transfer by conduction, convection, and radiation in your everyday life.

CHAPTER 9

WAVES AND SOUND

KEY TERMS

wavelength, amplitude, period, frequency, transverse wave, longitudinal wave, interference, reflection, refraction, diffraction, standing wave, Doppler effect, decibels

Waves are all around us every day, some in more familiar forms than others. If you are listening to music right now, the compressions generated by the motion of fabric panels in your speakers are sent out in front of the speaker in a pattern that strikes your eardrum; these compressions are called sound waves. If you are listening to the radio, sound waves from musical instruments or voices were converted into electrical signals that were sent out as a radio wave from a broadcast tower, received by the antenna on your radio, and then converted back into sound waves either by a speaker or by headphones. The light that illuminates the page you are reading is either reflected sunlight or incandescent or fluorescent lamplight.

In this chapter, we begin to discuss the fundamental properties of waves in their most familiar form: physical waves. These waves include water waves, waves on a string, and sound waves.

PHYSICS IN THE REAL WORLD

You are sitting in a rowboat in the middle of a lake, fishing, at 9 A.M. There haven't been any nibbles on your line all morning, and the lake is completely calm. In the distance, you hear the high-pitched whine of a speedboat. From its motion, you can see that the boat will come close to you. The speedboat skims across the lake and you realize that the underwater noises are probably scaring away any fish that may have been there. The boat approaches and passes you. As it passes, you notice that the pitch of the whine shifts lower. Once the boat has passed by you, it emits a lower pitch as it speeds away. The shift in pitch is due to the Doppler effect, a phenomenon that occurs when the source of waves (sound or light) is moving relative to you.

You see the ripples of the speedboat's wake approaching you, the crests about 2 m apart. As the waves pass you, you bob up and down in the water, a lot at first, and then less and less. You notice that you are bobbing up and down (one full cycle) once each second. The height of your bob is the amplitude of the wave passing you through the medium of the water, and the time it takes you to bob up and down is related to the wavelength of the water wave.

You are soon motionless again in your rowboat, waiting for a bite.

THE ANATOMY OF WAVES

Water waves are familiar to anyone who has been at the beach or on a lake. In fact, if you stand at the edge of a still lake and throw in a

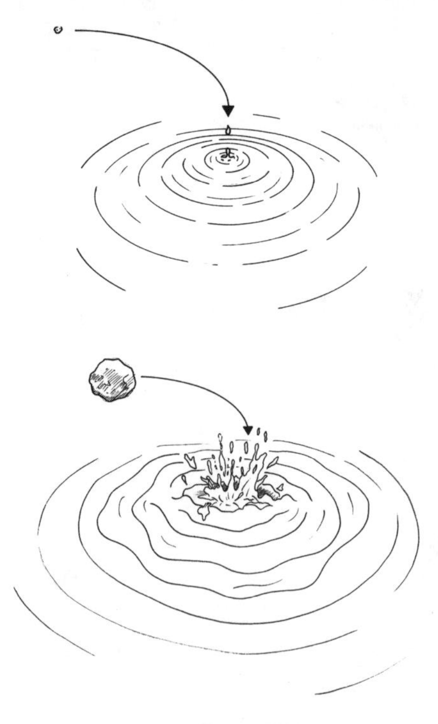

Figure 9.1
Waves in a Pond

rock, you will see circular waves moving out away from the point at which the rock entered the water. If you were to look at a profile of the surface of the water, you would notice that these waves have measurable characteristics; and it might not surprise you to find that a larger rock thrown into a lake produces a different wave than a smaller rock (see Figure 9.1).

Wavelength, Amplitude, and Wave Speed

Waves of all types have two fundamental characteristics: wavelength and amplitude (see Figure 9.2). The *wavelength* is the distance between two successive crests of the wave, and the *amplitude* is the "height" of the wave, measured as a displacement from some zero level. Wavelength is often abbreviated with the Greek letter lambda (λ). In our example of water waves and the rowboat, the wavelength would be the distance from one crest heading your way from the wake of the speedboat to the next. The amplitude of the wave would be the distance from the level of the calm lake to the highest point of the crest.

The *period* (T) of a wave is the time required for one complete cycle (crest to crest, or trough to trough) to move past a fixed point. In our example at the beginning of the chapter, the period of the water wave was 1 second, the time it took to bob up and down and back up again.

The *frequency* (f) of the wave measures how many complete cycles of the wave pass a fixed point in one second. Again referring to our simple example, the frequency was one cycle per second, or one hertz (Hz). The hertz is the unit of frequency and has units of 1/s. Frequency and period are reciprocals of each, or

$$f = 1/T$$

Waves also move through a medium at a measurable speed, and the wave speed can be determined by dividing the wavelength by the period, or

$$v = \lambda/T$$

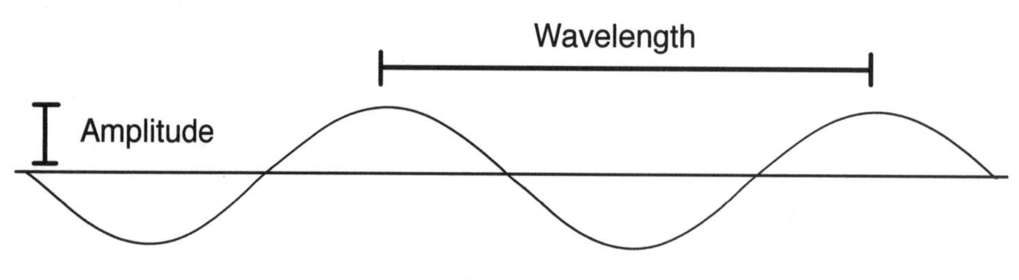

Figure 9.2
The Anatomy of a Wave

and since the frequency is the reciprocal of the period, the following equation is equivalent:

$$v = \lambda f$$

The speed of a wave generally depends on the nature of the medium through which it is moving. In other words, the speed of a water wave depends on some of the properties of water at a specific pressure and temperature. The speed of sound waves likewise depends on the density and temperature of the air through which waves are traveling.

EXPLORATION 9.1

What is the wave speed in the example of the person fishing in the rowboat at the beginning of the chapter?

We are told that the boat bobs up and down once each second, and that the separation between the wave crests is about 2 m. Using the formula for wave speed, we know that

$$v = \lambda f$$

Using $\lambda = 2$ m and f = 1/s, we have

$$v = (2 \text{ m})(1/\text{s}) = 2 \text{ m/s}$$

Transverse and Longitudinal Waves

There are two basic types of physical waves: transverse and longitudinal. Transverse waves are the type that we have been discussing thus far. In a *transverse wave,* the motion of the medium (water, a string) is perpendicular to the motion of the wave itself (see Figure 9.3a). In the case of the water wave, the water moves up and then down as the wave passes a point in the lake; the rowboat in the example moved only up and down while the waves moved past.

Longitudinal waves are waves in which the motion of the medium is in the same direction as the motion of the wave (see Figure 9.3b). Sound waves, for example, are longitudinal waves. Sound waves arise when the atmosphere is alternately compressed and stretched—for example, by the backward and forward motion of a speaker, or the clapping of your hands. In a longitudinal wave, there are alternations between low- and high-density regions; the low-density regions are called *rarefactions,* and the high-density regions are called *compressions.*

EXPLORATION 9.2

Have you ever been at a large sporting event (e.g., a football game or baseball game) when the fans started doing a "wave"? This type of a wave is a perfect example of a transverse wave. The wave moves around the stadium, while the "medium" (the fans) simply moves up and down. What do you do when the wave passes you? You stand (and scream) and then sit down.

This human wave has an amplitude (which is the difference between your sitting and standing height), a wavelength (which may be as large as several sections of seated fans), and even a wave velocity.

Think how hard it would be to get a longitudinal wave going in a stadium—perhaps at a soccer match it would work. Those are some rowdy fans.

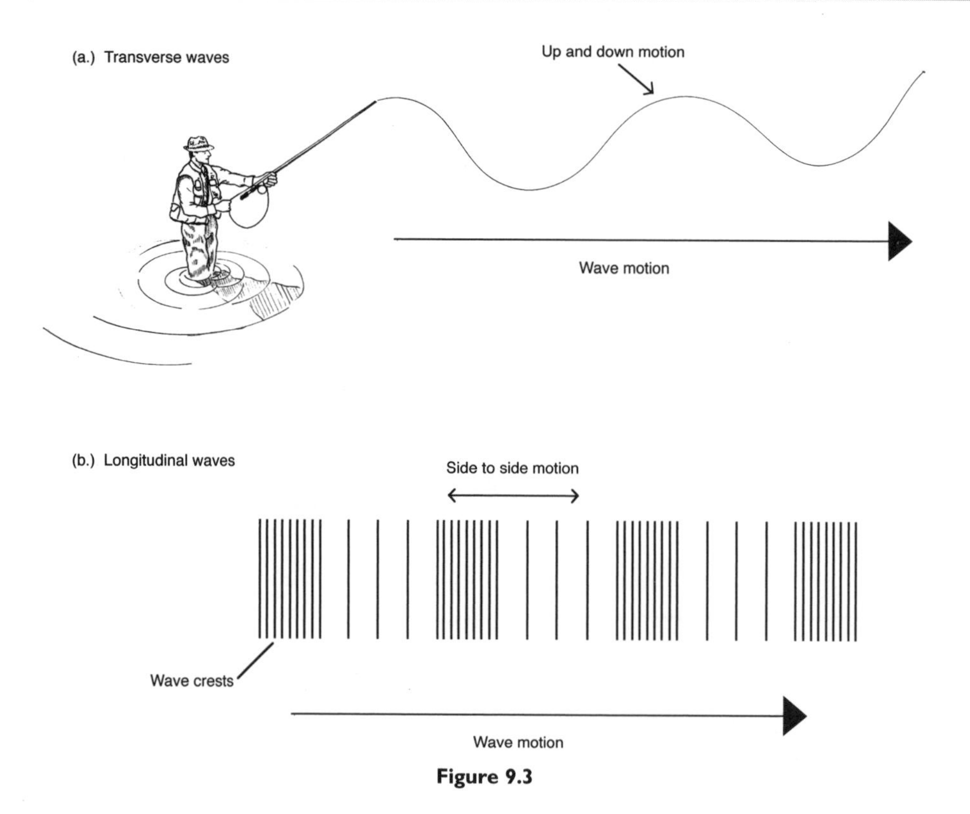

Figure 9.3

INTERFERENCE, REFLECTION, AND REFRACTION

Waves do all sorts of interesting things when they meet one another. Imagine that there are two water waves moving across the lake described at the beginning of the chapter. What happens when they meet? When two waves occupy the same physical space, the result is called *interference.*

How two waves interfere depends on where each wave is in its "cycle." There is a general rule that says that when two or more waves interfere, the combined wave will be the sum of the amplitudes of each of the interfering waves.

For a simple example, imagine two waves moving toward one another on a string (see Figure 9.4). If the two waves both have positive amplitudes when they meet, the amplitude will be the positive sum of the individual amplitudes. If they have equal amplitudes, for example, the interfering amplitude will be twice that of either of the individual waves. This situation is called *constructive interference.* If one wave has a positive amplitude and one has a negative amplitude when they meet, then the sum of the waves will be smaller than either of the individual waves. This situation is called *destructive interference.*

When waves encounter a barrier, there are several possible results. A wave may be diverted in the opposite direction. For example, a sound wave may strike a wall and bounce back so that you hear it as an echo. A light wave may strike a mirror and be directed in the opposite direction, which is called *reflection.*

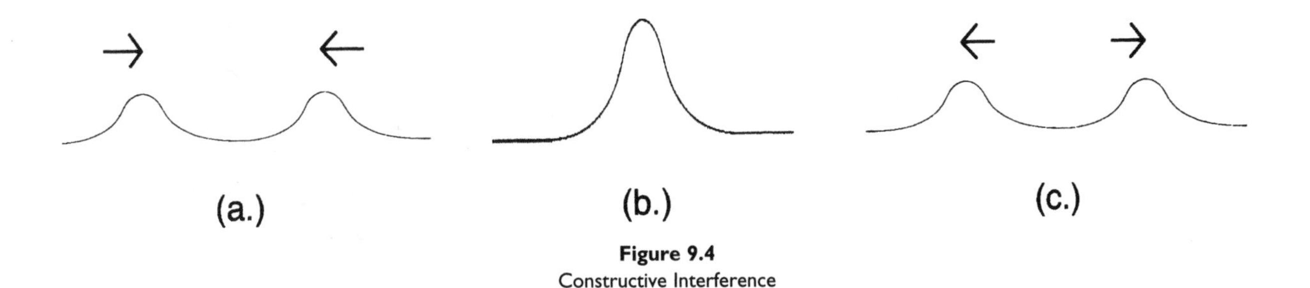

Figure 9.4
Constructive Interference

No boundary can reflect a wave entirely. Generally, part of a wave is reflected, and part of it is transmitted or absorbed. When a wave moves from one medium into another and strikes it at an angle, the wave often "bends" to move in a different direction because of a change in wave speed in the new medium, which is referred to as *refraction*.

When waves bend around the edge of an object, like water waves bending around a boat floating in a lake, the waves are *diffracted*. *Diffraction*, refraction, and reflection are all wave properties.

STANDING WAVES

If you attach one end of a rope to a wall and shake the other end up and down, you can generate a wave on a string. If you shake the rope up and down at just the right frequency, a wave of constant wavelength will appear "frozen" in the rope. This phenomenon is called a *standing wave*. *Standing waves* have fixed points that appear not to move at all (nodes) and moving points that appear to oscillate through the entire amplitude of the wave (antinodes) (see Figure 9.5).

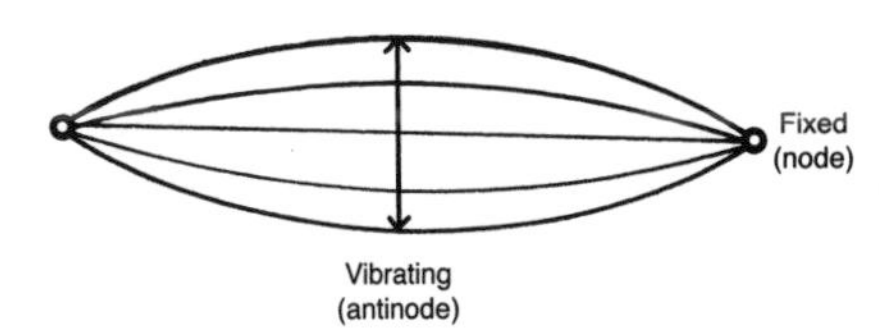

Figure 9.5 A Standing Wave

In the case in which both ends of the string are fixed, standing waves can be generated that are multiples of a half-wavelength (see Figure 9.6). The smallest possible standing wave would be a half-wavelength—then a full wavelength, then one and a half wavelengths, and so on. These types of standing waves are important in musical instruments in which both ends of the string are fixed (like the string on a violin). The frequencies at which large-amplitude standing waves are produced on a string are called *resonant frequencies*.

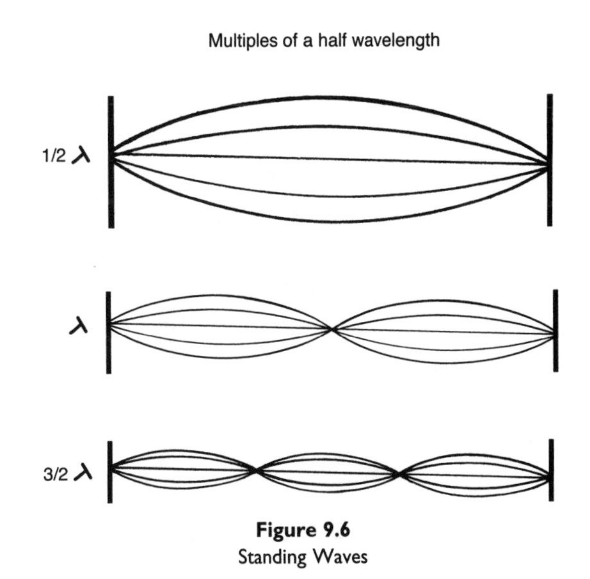

Figure 9.6
Standing Waves

SOUND WAVES

Sound waves are a type of longitudinal wave with which we are all familiar. We hear sounds around us every day, and our ears are sensitive detectors of sound waves, able to detect everything from a whisper to the audio barrage of a rock concert.

Sound waves require a medium to move through (Hollywood space explosions notwithstanding), and, generally, sound waves are transmitted through Earth's atmosphere. The compressions and rarefactions that move through the atmosphere are compressing the molecules of nitrogen and oxygen that are all around us.

The human ear, in general, is sensitive to sounds in a specific frequency range, from about 20 Hz to about 20,000 Hz. Sounds with frequencies less than 20 Hz are called *infrasonic,* and sounds with frequencies greater than 20,000 Hz are called *ultrasonic.*

Different animals have hearing that covers different ranges in frequency. A widely known example is that the hearing range of dogs extends to higher frequencies than our own. Sound waves that have a higher frequency are perceived to have a higher pitch (high notes).

EXPLORATION 9.3

If you have recently bought a piece of audio equipment (a stereo, a television, a CD player) and still have the technical specifications that are in the instruction booklet (usually near the back), find the frequency range over which your piece of equipment can produce sounds. You may not be surprised to find that it produces sounds in the range from 20 Hz to 20,000 Hz. Producing sounds outside that range would be a waste of time and money for a human customer.

EXPLORATION 9.4

On a warm summer night (30°C) you hear a clap of thunder about 3 seconds after you see the flash of lightning associated with it. How far away did the lightning strike?

To think about this problem, we must assume that we see the flash instantaneously. This is not a bad assumption, since the cloud is relatively close; but, as we will see in the next chapter, the assumption is not exactly true. If it takes 3 seconds for the sound to get to us, we can use the velocity of sound to see how far away the lightning struck.

Using a relationship for distance, time, and velocity we know that

$$\text{Distance} = \text{Velocity} \times \text{Time}$$

and

$$v = (331 + 0.6(30))\ \text{m/s} = 349\ \text{m/s}$$

so we can determine that the lightning must have struck at a distance of

$$\text{Distance} = 349\ \text{m/s} \times 3\ \text{s} = 1050\ \text{m, or about 1 km}$$

In general, each second you count off between the flash of lightning and the sound means that the strike was another kilometer away.

The speed of sound in a medium depends on the physical properties of the medium. The speed of sound in air (at a particular temperature) can be determined with the following formula:

$v = (331 + 0.6T)$ m/s

where T is the air temperature in degrees Celsius.

Like all waves, sound waves exhibit wave properties; that is, sound waves interfere, reflect, refract, and diffract. Sound waves reflect off hard surfaces (like buildings and canyon walls), and you can hear the reflection in the form of an echo. In fact, if you are standing in a canyon, you can measure the time it takes for the sound of your clapping to return to you. This time (and the velocity of sound) can be used to determine how far away the canyon wall is.

If you have ever sat at the edge of a lake on a summer night, you may have noticed that you can hear conversations on the far side of the lake, conversations that you would never hear during the day. This can happen if the air just above the lake is cooler—thus denser than the air farther up. The higher air density just above the lake causes sound waves to move more slowly there, and will cause sound waves to bend (refract) toward the surface of the lake, steering the sound waves toward you on the other side (see Figure 9.7). Those friends on the other side of the lake should be careful what they say about you at night!

Sound can also diffract (or bend) around obstacles, such as buildings or walls. If sound waves did not diffract, you would not be able to hear the sounds of someone walking down a hall at right angles to the hall you are in.

If you have ever gone swimming in a pool where music is playing, you know that most of the sound waves generated in the air are reflected at the surface of the pool. The volume of the music when your ears are in the water is low because although the amplitude is reduced owing to the transmission (refraction) of some of the sound at the pool surface, most of it is reflected back into the air. The opening scene of the movie "Saving Private Ryan" alternates between the hellish and loud sounds experienced by the troops above the water on D-Day, and the relative quiet of those troops that were submerged beneath the waves.

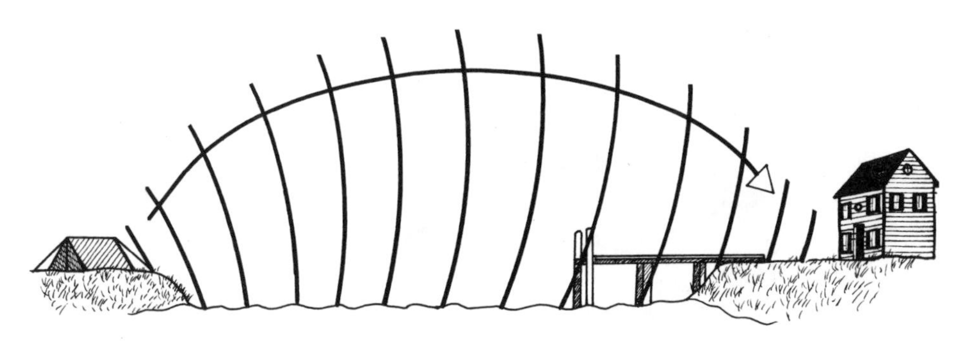

Sound waves bent (refracted) by the cool air above a lake.

Figure 9.7

The Doppler Effect

In the opening example, you were told that the pitch of the whine of the speedboat changed as the boat passed, moving from a high pitch to a low one. This change in pitch is an example of the *Doppler effect,* a variation in the detected frequency of a source of sound owing to the motion of the source relative to the observer.

In the case of the speedboat, its engine was emitting a whine of constant frequency, and since the whine was a sound wave, it was a longitudinal wave, moving away from the engine. The speedboat itself was in motion, though, so if we picture each compression in the sound wave as a crest, the wave crests were piling up in front of the boat—its direction of motion—and stretching out behind it (see Figure 9.8). Therefore, an observer in a position toward which the boat was moving would perceive the wave crests to be closer together (shorter wavelength and thus higher pitch); and an observer behind the boat would perceive the wave crests to be farther apart (longer wavelength and thus lower pitch).

The amount of the shift depends on the speed of the boat, of course. The faster the boat is moving, the greater the shift in pitch as it passes. Specifically, the frequency heard from a moving source by a stationary observer can be calculated to be

$$f_o = (v/v - v_s)f_s$$

where f_o is the frequency detected by the observer, v is the velocity of sound in the medium, v_s is the velocity of the source, and f_s is the emitted frequency (at the source).

Let's use our example of the boat coming across the lake. If we assume that the speed of the boat (v_s) is 10 m/s, the boat's engine emits a whine with a frequency of 256 Hz, and it is a 30°C day, then we can calculate the speed of sound in air to be

$$v = (331 + 0.6T)\ \text{m/s}$$

or

$$v = (331 + 0.6[30])\ \text{m/s}$$
$$v = (331 + 18) = 349\ \text{m/s}$$

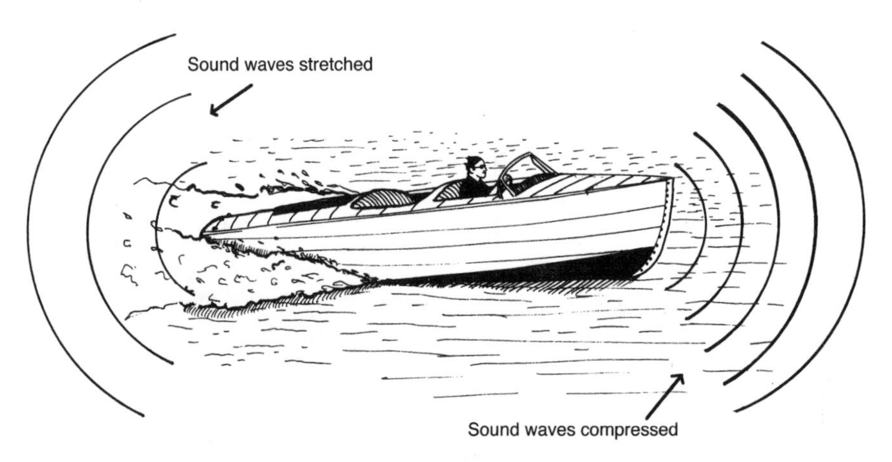

Figure 9.8

Using all this information, we can determine the frequency (pitch) of the detected sound to be

$$f_o = [v/(v - v_s)]f_s$$

or

$$f_o = [349 \text{ m/s}/(349 \text{ m/s} - 10 \text{ m/s})]256 \text{ Hz}$$

$$f_o = 264 \text{ Hz}$$

The observer will hear this higher frequency as a higher pitch. When the boat is moving away from the stationary observer, then the sign in the denominator becomes positive, yielding

$$f_o = [v/(v + v_s)]f_s$$

In this case, if the boat moves away with the same speed, the detected frequency will be lower. The solution to this second problem will be left for you to solve at the end of the chapter.

PHYSICS IN THE REAL WORLD

Have you ever heard a plane in the sky and looked up to see it, only to realize that you were looking in the wrong part of the sky? This can happen even if a plane is not moving faster than the speed of sound but is moving fast enough that the concentric sound waves that it is producing are stretched out behind it. When this happens, it is possible that a plane can pass above you silently before you hear it.

Shock Waves

Recall that the speedboat in the introductory example had a wake behind it. Why does a boat create a wake?

A wake results when the source of waves in a medium (in this case the speedboat making water waves) is moving faster than the speed that the waves can move in the medium (in this case water). If you recall, we determined that the wave velocity of the water waves was about 2 m/s, and in the preceding problem we said that the speedboat was moving at 10 m/s. Thus, because the speedboat is moving faster than the speed of the waves, it generates a wake. The wake of a boat is a shock wave, similar to another shock wave you might have heard—a sonic boom.

A *sonic boom* results when an airplane moves through the air at a velocity faster than the speed of sound in air. The plane is basically outrunning the sound waves it is producing, and again, it creates a wake—in this case, a wake of sound. When this wake passes you, you hear a sonic boom.

THE SOUND INTENSITY SCALE

We end the chapter with a little bit of practical information about the intensity of sound. It shouldn't surprise you to know that the amplitude of a sound wave determines its "volume" or intensity. A large-amplitude sound wave will sound louder than a sound wave with a smaller amplitude; and the farther you get from the source of a sound, the quieter it becomes. In fact, the intensity of a sound wave decreases as the distance squared. Why is this so?

The intensity of a sound decreases rapidly because sound waves move away from a sound source (imagine a tiny bell) in all directions. The energy of the sound wave

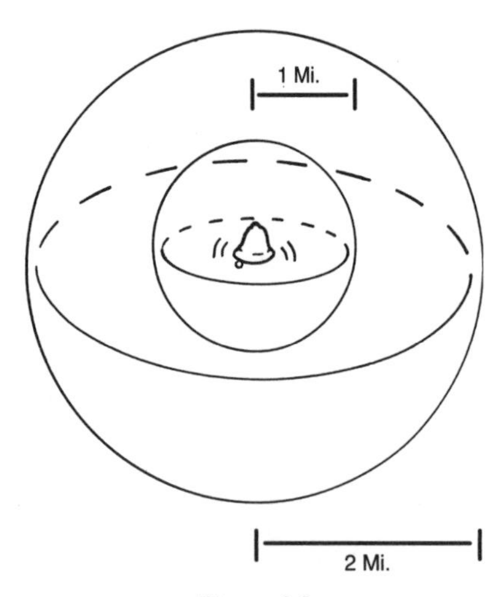

Figure 9.9

decreases with the area through which it passes. Thus, if you are 1 m from the tiny bell, the energy passes through a sphere with a 1-m radius. If you are 2 m from the tiny bell, the energy passes through a sphere with a 2-m radius (see Figure 9.9). Because the energy passes through the surface of the sphere, and surface area is determined by

$$\text{Surface area of a sphere} = 4\pi r^2$$

the intensity decreases with the *square* of the distance from the sound source. The energy from the sound wave is diluted because it passes through increasingly larger areas centered on the tiny bell (see Figure 9.9).

The relative loudness of sounds is determined on a *logarithmic scale* and is measured in *decibels* (dB), named in honor of Alexander Graham Bell (1847–1922). On the decibel intensity scale, a sound that is 10 dB is 10 times as loud as a 0-dB sound. A sound that is 20 dB is 10 times louder than a 10-dB sound and 100 times louder than a 0-dB sound. A sound with a loudness of 50 decibels is 100,000 times as loud as a 0-dB sound.

Source	Loudness (dB)
faintest audible sound	0
rustling of leaves	8
whisper	10–20
average home	20–30
automobile	40–50
ordinary conversation	50–60
heavy street traffic	70–80
riveting gun	90–100
thunder	100
amplified rock music	115

Table 9.1 Sound Intensity Levels

The human ear is damaged when intensity approaches about 85 dB. Table 9.1 gives the decibel values of some common sounds and should give you an idea of why those construction workers with jackhammers have plugs in their ears. The plugs (made of wax or foam) absorb or reflect the sound, reducing the intensity that is transmitted to the eardrum to a bearable—and safe—level.

PROBLEMS

9.1 If the wave crests were 3 m apart in the introductory rowboat example and the period of the waves was the same, what would be the wave velocity?

9.2 Waves can constructively or destructively interfere when they meet as described in the chapter. What would happen in a stadium if two "waves" collided? What is different about a human "wave"?

9.3 It is possible for two waves to have total destructive interference, meaning that the amplitude when they meet is zero. Draw a sketch of how this might happen with two waves on a string.

9.4 At a stereo store, a salesman tells you that the expensive new amplifier can produce frequencies up to 50,000 Hz, which is why it carries a premium price. What should your response be?

9.5 What is the speed of sound in superheated 100°C air?

9.6 Why does temperature have an effect on the speed of sound in a medium?

9.7 What is the frequency of the speedboat whine as it speeds *away* from you across the lake? Is it a lower or a higher frequency? (This is the second part of the example problem in the section on the Doppler effect)

9.8 How fast does an airplane have to be moving to produce a sonic boom on a day when the temperature is 30°C?

9.9 You may have noticed when you fly in an airplane that the window gets very cold, indicating that as you move higher in Earth's atmosphere, it gets colder. If this is true, what happens to the speed of sound high in Earth's atmosphere? Based on your answer, what does this mean in terms of the production of a sonic boom?

9.10 How much louder is a normal conversation than whispering?

LIGHT AND OPTICS

KEY TERMS

speed of light, illumination, light rays, law of reflection, refraction, index of refraction, virtual image, real image, convex mirror, concave mirror, lenses

If you are sitting in a chair or at a desk right now, look out the window. What do you see? Trees and a yard, and beyond that a road? An adjacent building? The images of everything that you can see out your window—and everything that you see in the room where you are sitting—come to you in the form of light. Rays of light from the Sun are reflected from all the surfaces that you see, and these reflected rays enter your eye. The lens in your eye focuses (through refraction) the rays of light onto your retina, where the energy of the light is deposited and converted by your brain into both intensity and color. If you were to look out the same window at night, you might see a very different scene. If you live out in the country, you might not see much at all (owing to the absence of a source of illumination). If you live in the middle of a city, streetlights might illuminate the scene. These streetlights would become the source of light rays that reflect off surfaces and enter the lens of your eye.

In this chapter and the following one, we examine the properties of light. First, we study light as streams of particles that move in straight lines, reflecting off surfaces or bending through materials, as they travel. In the next chapter we will examine in some detail the wave properties of light.

WHAT IS LIGHT?

What exactly is light? It is all around us, and yet it is rather difficult to define. Light clearly emanates from some objects (like the Sun) and not from others, as we can perceive when we block the path of the light. The shadow of your body on the sidewalk on a sunny day is evidence that rays of light originate from the Sun. The shadow of your body on the same sidewalk at night is evidence that streetlights also can produce light.

Light carries information about the object from which it is reflected. In the simplest sense, light rays reflecting off an object into your eye tell you that the object is there. Light rays bouncing off an object—or traveling through a substance—are changed slightly, and these changes tell us even more about the object involved, as we will discuss in the next chapter.

As early as the sixteenth and seventeenth centuries, scientists were beginning to understand that light had some of the properties generally associated with waves. Sir Isaac Newton considered this possibility but concluded that light rays were akin to streams of particles that reflected off surfaces and bent (or refracted) through materials like glass when they struck them at an angle.

PHYSICS IN THE REAL WORLD

Some objects that seem luminous themselves are simply reflecting light. The Moon is a perfect example—it generates no energy or light on its own but glows only in the reflected light of the Sun. When Earth is on the fully illuminated side of the Moon, we see a full moon. When, later in its orbit, the Moon has the Sun at its back, we see the Moon as dark, or as a mere sliver (new moon). Without the Sun, the Moon would be dark indeed, as would our entire solar system.

In the centuries after Newton, it became clear that light had wave properties. In the early twentieth century, the work of Albert Einstein again suggested that light behaved in some ways like a particle and carried energy in small packets, or quanta. We will discuss Einstein's proposal and work in detail in Chapter 15, "The Atom and Quantum Mechanics." For now, we simply state that observations indicate that light has the properties of both waves and particles and that the *wave–particle duality*, as it is called, has no effect on the reflecting and bending of light that we discuss in this chapter.

The Speed of Light

In the previous chapter we discussed the wave velocity of water waves and sound waves and stated that the velocity of a wave depends on the medium through which it travels. Although light waves do travel at very high velocity, they do not travel instantaneously (despite our assumption in the thunder and lightning example in the last chapter). Light takes time to travel from one place to another, but since it travels so rapidly, it is difficult to measure the velocity of light.

One might imagine two observers on widely separated hilltops, each with a lantern, and the observers could agree to uncover the lanterns at a specific time. One observer could try to measure the delay between the uncovering of the lantern on the distant hillside (at a preordained time) and the perception of the uncovering to determine the velocity of light. This experiment was tried many times—unsuccessfully, because light travels too fast for this experiment to work.

The first calculation of the speed of light was made by the Danish astronomer Olaus Roemer (1644–1710)—quite by mistake. He was working on the orbits of the moons of Jupiter with the thought that precise periods of the moons' orbits might serve as a celestial timepiece. Roemer was surprised to notice that the moons appeared at different times than expected when Earth was in different points in its orbit (effectively at different distances from Jupiter). He calculated the speed of light by timing the appearance of one of the moons of Jupiter (as it emerged from behind the planet in its orbit) at different times in Earth's orbit. Using different points in Earth's orbit put the planets at different distances and allowed him to measure the different travel times. With a distance and a time, Roemer was able to calculate a velocity. His calculation of the speed of light (abbreviated with the letter c) was very close to what we now know to be the correct speed of light (over 180,000 miles per second). The high velocity—carrying light from the Sun to Earth, for example, in just 8 minutes—explains why it was difficult to make the two-lanterns experiment work.

In modern laboratories, the *speed of light* is calculated from the time it takes a beam of light to cover a known distance. Because light travels about 1000 ft in one-millionth of a second, the need for accurate timing is obvious. The best current measurements of c (in a vacuum) is

c = 183,310 mi/s

or

c = 299,776 km/s

EXPLORATION 10.1

If light travels at 183,310 mi/s, how long would it take light to complete a marathon?

The marathon distance is 26.2 mi. Therefore, the time required is

t = distance / velocity
t = 26.2 mi / 183,210 mi/s
t = 0.00014 s

or less than a millisecond!

Light is significantly unlike the other types of waves we have studied so far. Unlike sound waves, or water waves, or human waves at a football game, light does not require a medium through which to travel. In the nineteenth century, physicists tried to identify such a medium, and even had a name for it: the *ether*. Measurements were attempted to determine how fast Earth was traveling through the hypothesized ether. In the middle of that century, James Clerk Maxwell (1831–1879) proposed instead that propagation, or movement of light waves, is the result of oscillations in what are called *fields*—the electric and magnetic fields in particular. Therefore, light can travel through the vacuum of space (at the speed of light), since space is filled with electric and magnetic fields, not with ether.

When light travels through regions other than a vacuum, its speed is measured to be slower. For example, in air, light is slowed by just 0.03 percent; in water it is slowed by 25 percent; and in glass by 35 percent. This change in the velocity of light as it moves from one substance to another is responsible for the refraction or bending of light and makes possible all optical equipment, from eyeglasses to microscopes and telescopes.

LIGHT AND ILLUMINATION

Sources of Light

On Earth, the Sun is the main source of natural illumination. It should come as no surprise that early on, humans began to devise forms of artificial illumination. Like campers today, ancient humans must have used firelight to illuminate their surroundings at night.

An icon of modern society, the electric lightbulb is formally known as a *filament lamp*. In such a lamp, light is produced when a current of electricity passes through a very fine tungsten wire, or filament, placed inside a glass bulb. The bulb must either be evacuated or contain an inert gas (like nitrogen) so that the filament will not ignite when it heats up, as the current passing through the tungsten filament raises its temperature to about 2500°C. If oxygen is present in the bulb, the filament will ignite, and the bulb will "burn out." In some lightbulbs, a white coating inside the bulb scatters the light and "softens" it. In other bulbs the glass is clear, and the light is more intense.

A *carbon arc* lamp is a very intense source of light used in theater movie projectors and in searchlights. Two carbon rods are connected to an electric battery or generator. The tips of the rods are brought into contact and then drawn apart a short distance, producing a region that carries electric current across the gap. Most of the light comes from the glowing tips of the carbon rods, which reach a temperature of 3000°C to 3500°C. The higher temperature makes the bulb far brighter than an ordinary lightbulb.

What are sometimes generically called "neon lights" or "neon signs" are constructed from bent tubes of glass filled with gaseous elements or compounds. A current of electricity flows through the gas or vapor and generates light of various colors. Tubes filled with actual neon gas, for example, emit an intense red light; hydrogen-filled tubes emit a subtle pink light; and helium gas gives off a light blue glow. The tubes may also contain easily vaporized substances such as sodium or mercury.

In fluorescent lamps, the production of light is more indirect. Electricity passes through a mixture of argon gas and mercury vapor, generating ultraviolet light. Ultraviolet light is not visible to the human eye, but its energy is absorbed by a chemical coating on the inside of the tube, causing it to glow.

Discharge lamps produce light by a combined action of a gas and a filament; that is, part of the light arises from the filament itself, and part arises from the gas in the bulb giving off its own glow. Sodium vapor lamps used on expressways, halogen automobile headlights, and halogen lights used for home lighting are examples of this kind of bulb.

Other light sources include light-emitting diodes (LEDs), arrays of which can produce very intense light sources with low power consumption. Some streetlights and even consumer electronics (flashlights) now use LEDs as a source of illumination.

The Intensity of Light

The intrinsic brightness of a light source is specified by a quantity called *luminous intensity*, which is measured in a unit called a *standard candle* or *candela*, abbreviated *cd*. As the name suggests, this unit dates to a time when a wax candle was the standard source of light. Astronomers even today apply the term *standard candle* to an astronomical object whose luminosity is known with confidence. They compare this luminosity with the object's apparent brightness to determine the distance to the galaxy in which the object resides. One candle or candela is equal to 1/60 of the luminous intensity per square centimeter of a perfect emitter of radiation (known as a *blackbody*) at a specific temperature (the temperature of the solidification of the element platinum [2046 K]).

Luminous flux is the rate of flow of energy carried by light. The SI unit of luminous flux is the *lumen*. Imagine a 1-candela (1-cd) source located at the center of an imaginary sphere. All the luminous intensity of the 1-cd source flows through the sphere. A lumen is defined as the fraction of luminous intensity from a 1-cd source that flows through an area that is $1/4\pi$ of the surface of an imaginary sphere centered on the light source. This fraction of the imaginary sphere is also called a *solid angle* equal to one *steradian*. The surface of a sphere covers a solid angle of 4π steradians.

Illumination is defined as the amount of luminous intensity falling on a given area. Like sound, light spreads out in waves from its source. The farther an object is from the source of light, the larger the area over which a given amount of light will spread. This area increases as the square of the distance (as described for sound in the previous chapter—see Figure 10.1). At twice the distance, a given amount of radiant energy will be spread over four times the area. At three times the distance, it will be spread over nine times the area, and so on.

This relationship means that the illumination falling on each unit area will vary inversely as the square of the distance from its source. The illumination of any surface that is held perpendicular to incoming rays of light will depend directly on the luminous intensity of the source, and inversely on the square of the distance to the source. A point source with a luminous intensity of 1 candle (or 1 cd) will result in a luminance of 1 *foot-candle* (ft-c) at a distance of 1 ft and a luminance of 1 *meter-candle* at a distance of 1 m. Lighting engineers and landscape architects who are trying to achieve a specific level of illumination for interior and exterior areas use the terms *foot-candle* and *meter-candle.*

The relationship among illumination, luminous intensity, and distance can be given by:

$$E = C/d^2$$

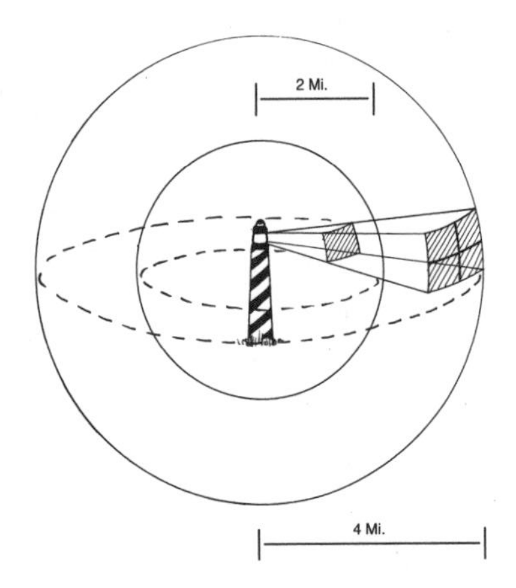

Figure 10.1

where luminous intensity (C) is measured in candles and distance (d), in feet. Although illumination (E) is sometimes expressed in *foot-candles,* inspection of the units in the preceding equation shows that the units actually are candles per square foot or candles per square meter.

The unit of illumination in more common use in the sciences is the *lux,* a unit of illumination equal to 1 lumen per square meter. A lux is the equivalent of about 0.1 ft-c (0.0929 ft-c to be exact). The nerves in the human eye can be stimulated by as little as a *ten-billionth of a foot-candle,* or *one hundred-billionth of a lux,* which is the equivalent of the illumination produced by a single candle nearly 20 miles away.

The intensity of a lamp can be measured by comparing the illumination it produces (in lux) with that from a standard lamp by means

EXPLORATION 10.2

Try putting your knowledge to use by computing how far you need to place a lamp with a 60-cd bulb to provide 15 ft-c (candles per square foot) of illumination on the book you are reading.

We use the relationship $E = C/d^2$

$$15 \text{ cd/ft}^2 = 60 \text{ cd}/d^2$$

or

$$d^2 = 60 \text{ cd/ft}^2/15 \text{ cd} = 4 \text{ ft}^2$$

$$d = 2 \text{ ft}$$

Therefore, you need to place the lamp at a distance of 2 ft in order to have a 15 ft-c level of illumination.

of a device called a *photometer*—photographers often use photometers when calculating exposure times. Engineers and architects also have devices that can directly measure the total illumination of any surface.

RAYS OF LIGHT

When people began to study light in detail, one of the first things they noticed was that it travels in a straight line. Light seems to take the most efficient (shortest) path from point *A* to point *B*. Ancients knew this as well. They constructed buildings (like the *kivas* in New Mexico) so that when the Sun was located at a particular position on the horizon, sunlight would enter through a small opening and illuminate a particular part of the interior structure.

When you stand outside on a sunny day, your shadow is cast behind you in straight lines from the Sun to the ground. When driving at night, you have surely noticed that your headlights beam straight ahead, making it hard to see around a curve. Light appears to travel in straight lines that we call *light rays*.

As a child, you may have made shadow puppets on the wall. All you needed was a light bulb, a darkened room, and your hand (see Figure 10.2). Imagine straight lines drawn

Figure 10.2

EXPLORATION 10.3

You can easily make what is called a pinhole camera, which utilizes an important property of light: it travels in straight lines. Cut out one end of a small cardboard box and cover this end with a piece of translucent paper (tracing paper or wax paper are good choices). Make a clean hole in the center of the opposite end of the box with a thick needle (see Figure 10.3). In a darkened room, a candle flame or a light bulb placed a few feet in front of the hole will produce a bright, inverted image on the wax paper. Move the candle back and forth, and notice that the closer you bring the object to the camera, the larger its image will be.

Surprisingly good pictures can be obtained with a pinhole camera by putting a photographic film in place of the screen. In fact, the device that you have just made is very similar to the first cameras made in the nineteenth century; however, because only a small amount of light can come through the pinhole, exposures of photographic film are inconveniently long.

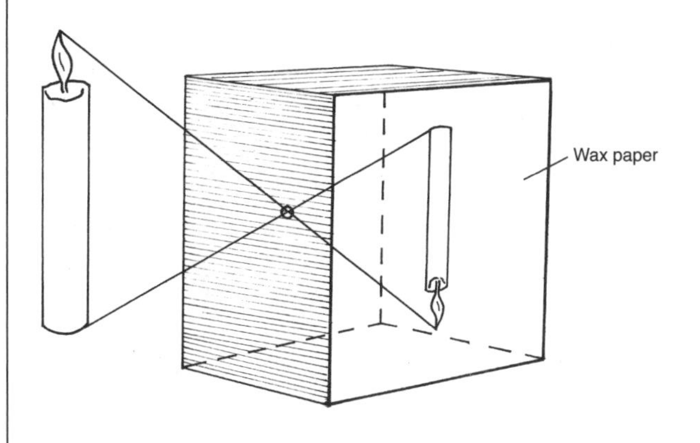

Figure 10.3
A Camera Obscura, or Simple Camera

from the light source touching the edge of your hand, and then striking the wall. All the rays that strike your hand are blocked, casting a shadow in the shape of your hand. The lines emanating from the bulb represent light rays. These rays travel outward from the source in all directions. For you to see an object, light rays coming from it must enter your eye.

REFLECTION AND REFRACTION

Reflection

When light strikes a barrier—or moves from one medium to another—part of the ray is reflected and part of it is transmitted. Depending on the properties of the material, more or less of the light may be reflected, or sent back in the direction from which it came. Some materials (mirrors) are more reflective than others (flat black paint) to visible light.

When light is reflected from a plane (flat) mirror, the incoming ray of light, called the *incident ray,* and the reflected ray of light, called the *reflected ray,* can be measured with respect to the *normal,* a line perpendicular to (at right angles to) the plane surface (see Figure 10.4). When any light ray strikes a plane, the angle of incidence (measured from the normal to the incident ray) always equals the angle of reflection (measured from the normal to the reflected ray), on the other side of the normal. This relationship is known as the *law of reflection,* and it can be used to determine the image that will be formed by any reflective surface, whether it is flat or curved.

As we have mentioned, light rays move out in all directions from an imaginary point source of light; however, if you imagine the rays of light at a considerable distance from

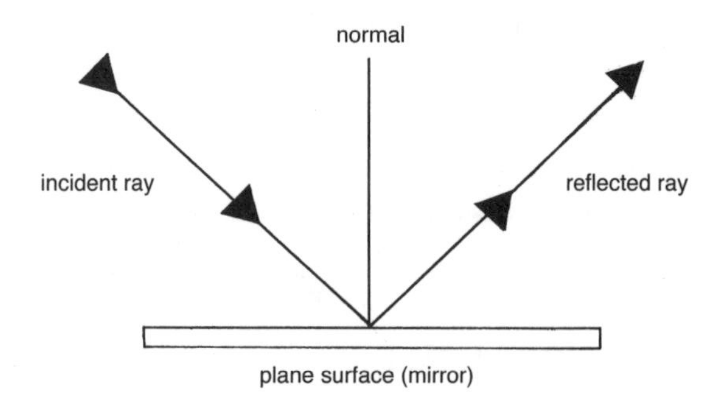

Figure 10.4

the source, they begin to approximate parallel rays of light, as shown in Figure 10.5. The rays of light that reach us from the Sun are a good example. The Sun is so far from us (1.5×10^{11} m), that when its radial rays reach the surface of Earth, they are for all practical purposes parallel rays of light.

Types of Reflection

If parallel rays of light hit a perfectly flat surface, the law of reflection tells us the rays will also be parallel after reflection (see Figure 10.6a). If rays of light strike a rough or irregular surface (Figure 10.6b), however, the angles of incidence and reflection will be equal at each individual point on the surface, but the surface will reflect incoming parallel rays in many different directions. Reflection from a rough or irregular surface is called *diffuse reflection.*

A highly polished silver surface can reflect up to 95 percent of the light that falls on it. An ordinary mirror, consisting of a sheet of glass silvered on its back, reflects about 90 percent of the incident light. Even without a silver backing, a smooth surface like a sheet of glass reflects some light, though the amount of light reflected is generally less than 10 percent.

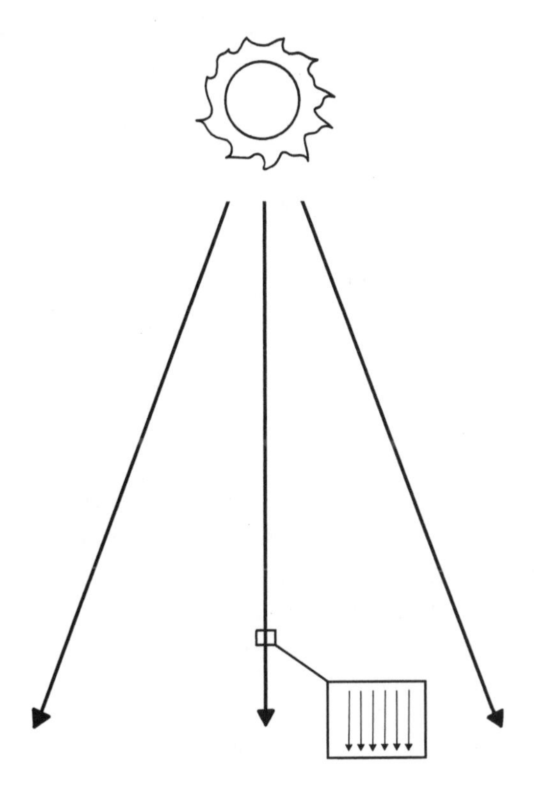

Figure 10.5

At increased angles of incidence (called *grazing incidence)*, a larger percentage of incident light is reflected. This phenomenon explains why the Sun's reflection in a lake is not particularly bright when the sun is overhead (most of the light is transmitted and very little is reflected) but can be too intense to look at when the Sun is low in the sky and about to set (most of the light at this grazing incidence is reflected from the surface of the water).

EXPLORATION 10.4

Any smooth surface can function as a mirror. The rougher the surface, the less light it will reflect in a coherent manner. Find different flat materials around your house, and hold them under a lamp. Which types of materials form the brightest image in reflection? Which form little or no image? What do these differences tell you about the smoothness of the reflecting surface?

Refraction

A mirror functions by bouncing back light striking its surface into the space from which it comes. All surfaces reflect some light, but if the surface is made of a transparent material, some of the light passes through into the second medium. If a light ray strikes at an angle to the normal, the ray will change its direction as it passes into the second material. This change in the direction of a ray is called *refraction.*

If a ray of light passes from a less dense substance (e.g., air) into a more dense substance

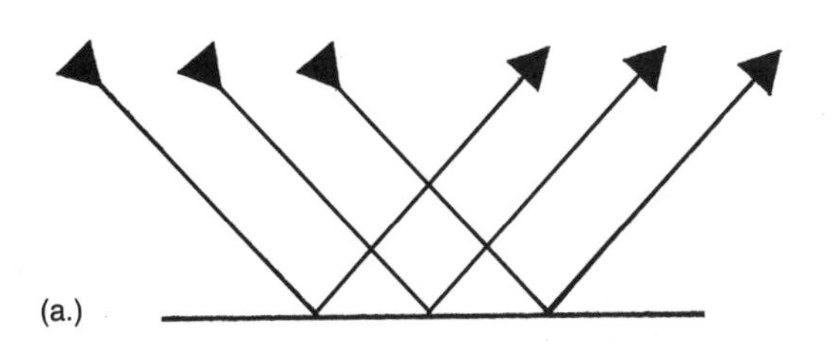

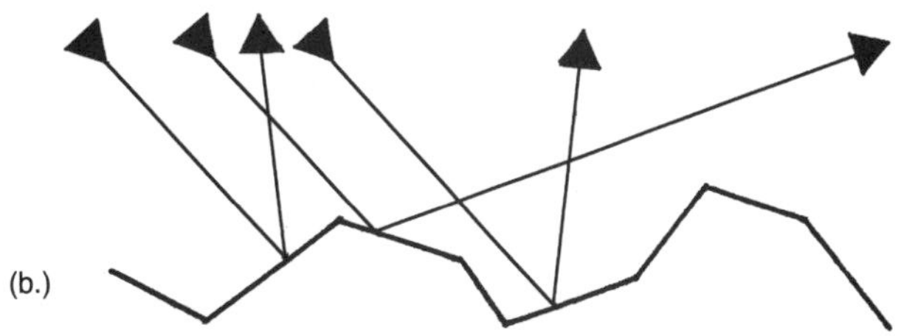

Figure 10.6

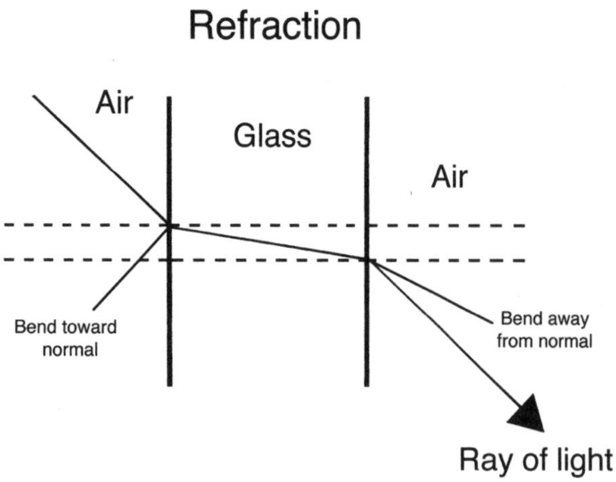

Refraction

Air

Glass

Air

Bend toward normal

Bend away from normal

Ray of light

Figure 10.7

(e.g., glass), the ray will bend *toward* the normal. If the ray passes from a denser to a less dense substance, the ray will bend *away from* the normal. Figure 10.7 shows these two effects.

> **EXPLORATION 10.5**
>
> Have you ever looked down at your legs while standing in shallow water? Or looked at a spoon standing in a glass of water? In each case, the object in the water (your legs or the spoon) appears to bend at the surface of the water.
>
> This apparent bending occurs because the light rays that reflect off the object under the water are refracted as they move from the water into the air (away from the normal). In general, a submerged object will appear to be closer to the surface than it actually is. Figure 10.8 shows a light ray diagram that indicates why this happens. At home put a pencil or knife into a glass of water to see if you can observe this effect.

When light passes completely through a piece of glass that has parallel sides, the rays bend, or refract, at each of the surfaces. As the rays enter the glass, they bend toward the normal, and as they leave the glass they bend away from the normal by the same amount.

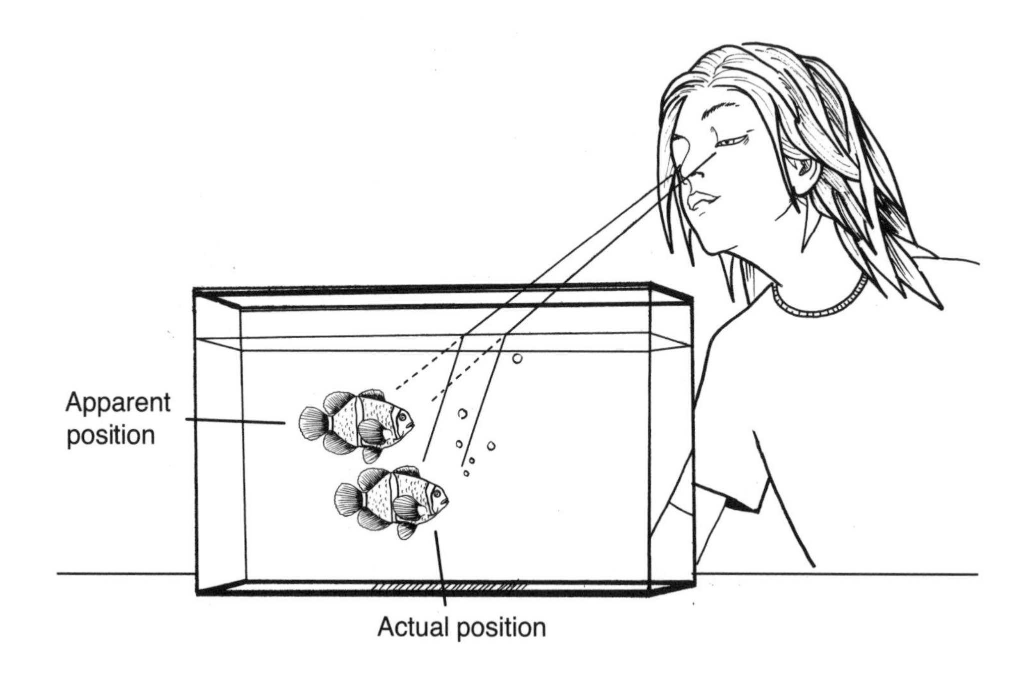

Figure 10.8

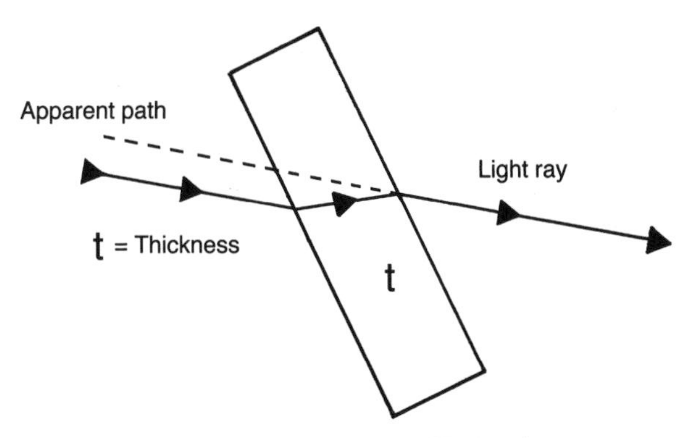

Figure 10.9 An Example of Refraction

Figure 10.9 shows this effect in a piece of glass with thickness *t*.

The light that is transmitted through the glass (some small fraction is lost to internal reflections at each of the boundaries) is parallel to its original course but offset to one side. You can easily observe this phenomenon yourself. Simply hold the surface of a piece of glass at an angle to your line of sight and hold it so that you see part of the pencil through the glass, and part of the pencil directly. The part of the pencil behind the glass will appear to be displaced to one side.

Index of Refraction

The ratio of the speed of light in a vacuum to its speed in a given material is called the *index of refraction* of the material. This quantity, represented by the symbol n, determines how much rays are bent when they enter a material and is determined to be

$$n_s = c/v_s$$

where n_s is the index of refraction of a substance and v_s is the speed of light in the same substance. Because the speed of light is less than c, unless light is traveling in a vacuum, the index of refraction of all substances is a number greater than 1.

An interesting example of refraction and reflection can be observed when light is traveling toward the surface of water from below. Figure 10.10 shows a sketch of an underwater spotlight that can be inclined at various angles to the surface. As the rays of light from the lamp are inclined more parallel to the surface of the water, the beam emerges closer

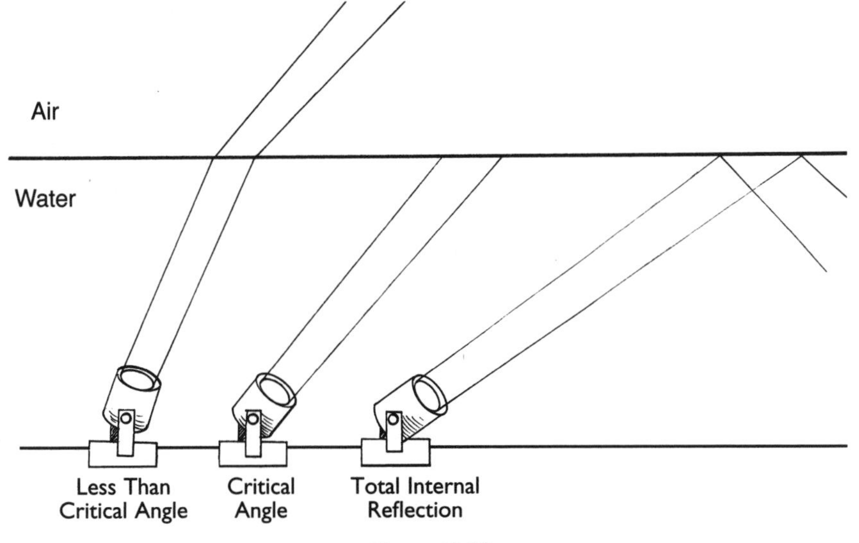

Figure 10.10

and closer to the surface. At some *critical angle,* the outgoing beam does not emerge at all but skims the surface of the water. If the angle is made any greater than the critical angle, the beam is completely reflected back into the water. This phenomenon is called *total internal reflection.*

Materials with a higher index of refraction have a larger range of incident angles through which total internal reflection occurs. Fiber optic cables, for example, operate on this principle. These are thin, transparent glass or plastic fibers that are enclosed in a material with a lower index of refraction. Pulsed laser beams travel very effectively through fiber optic cables, because the beam experiences total internal reflection as long as the cable does not have any sharp bends.

Substance	Index of Refraction (n)
vacuum	1.00
air	1.0003
ice	1.31
water	1.33
gasoline	1.38
olive oil	1.47
glass, ordinary	1.5
glass, dense optical	1.6–1.9
diamond	2.42

Table 10.1 Index of Refraction of Various Substances

MIRRORS AND LENSES

Mirrors

Imagine you are looking at a candle held in front of a flat (plane) mirror. This setup is illustrated in Figure 10.11. The rays of light emanating from the candle strike the mirror and reflect off its surface at an angle of incidence equal to the angle of reflection for each ray. If you are looking into the mirror, you will perceive these reflected rays as coming from behind the mirror. Your eyes will trace the rays back to their apparent origin, and you will see an image of the candle in the mirror. If the mirror is flat, the image will have the same size as the candle, and it will appear to be as far behind the mirror as the real candle is in front of it.

We sometimes talk about looking *into* a mirror for this reason. There appears to be an entire parallel world beyond the surface of the mirror. In Lewis Carroll's *Through the Looking Glass,* Alice enters that strange reflected world

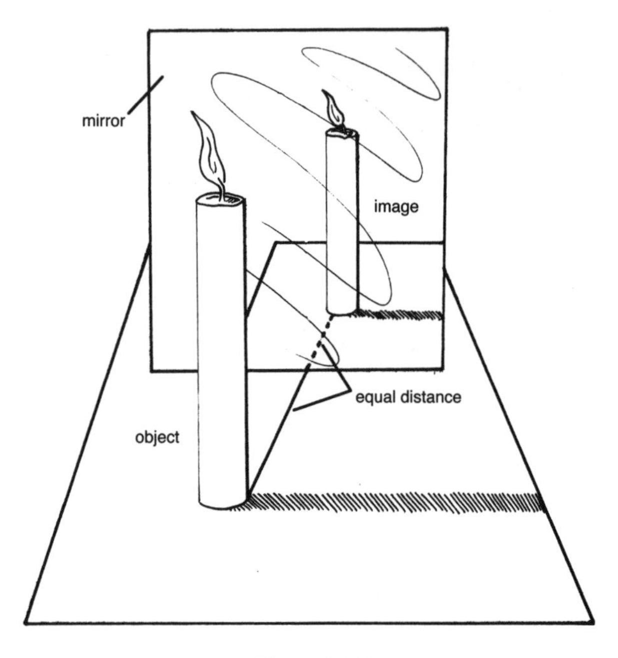

Figure 10.11
Reflection in a Plane Mirror

by stepping through the plane of the mirror. Of course, nothing is really going on behind the reflecting surface of a mirror, and the candle that hovers out of reach on the other side of the mirror is called an *image* or a *virtual image.* A virtual image is any image of an object that cannot be focused, because the light rays only appear to originate from it.

When you look at your own virtual image in a plane mirror, you know that right and left are reversed. If you wink your right eye, the left eye of the image winks back.

A mirror can also create a *real image,* that is, an image that can be focused (like the image in the pinhole camera in Exploration 10.1). Mirrors that have curved surfaces can be used to create real images. Modern research telescopes use curved mirrors to focus the light of distant stars and galaxies (see the final section).

EXPLORATION 10.6

If you have a mirror in your house, stand close to the mirror and look to the left and right. You will notice that you appear to be able to see objects that are not located in front of the mirror. That is, you can see an object to your right (in the room) over to the right in the mirror. How can this be? How can an object that is not in front of the mirror form an image in the mirror?

Mirrors operate on the principle of reflections of rays, so the rays of light from an object to your right (for example) will bounce off the surface of the mirror and into your eye (see the diagram). Your brain traces the reflected ray back in a straight line and sees a virtual image of the object in the mirror, the same size and distance from the mirror as the real object.

If the center of a mirror bulges toward you, the observer, it is a *convex mirror.* If the center bulges away from you, it is a *concave mirror.*

You may have noticed that the side-view mirror on the passenger side of a car says "OBJECTS IN MIRROR ARE CLOSER THAN THEY APPEAR." This statement alone tells you that the mirror on that side of the car cannot be a plane (flat) mirror. If it were flat, then objects would appear to be at their correct distance from the mirror, as described previously. Utilizing a curved surface (convex in this case) accomplishes another goal: it creates a virtual image of the area behind your car that is reduced in size—a virtual, reduced image. This allows you to see more of the area behind your car in a small mirror. The drawback is that you have to be reminded that the size of the cars in the image is reduced slightly, making them look farther away.

Lenses

Lenses are curved pieces of glass familiar to us in many everyday objects, such as eyeglasses, magnifying glasses, projectors, and cameras. The bending of light rays in lenses is the result of refraction at the interface between two materials. Lenses function because of refraction, and the extent of the curvature of the two surfaces of a lens determines how much light is bent. Parallel light rays can be bent to bring them together at a focus.

Lenses are generally made of glass or plastic. To make a lens with specific properties, a technician grinds the two facing surfaces with a spherical form.

In analogy with mirrors, lenses that are thicker at the center are called *converging lenses,* and lenses thicker at the edges are called *diverging*

EXPLORATION 10.7

Go to your kitchen and take a spoon out of the drawer. The cupped side of the spoon is a concave mirror, and the opposite side of the spoon is a convex mirror.

Look at yourself in the convex side of the spoon. What you see is a reduced, virtual image of the room around you. The image is also distorted because (unless you have a very unusual spoon) your spoon is not spherical. As you look at yourself in the spoon, rotate it. The distortion is similar to the visual defect known as *astigmatism* (see the final section, on eyeglasses). No matter how close you get to the convex side of the spoon, the image does not change; it always produces an upright, reduced image.

Now, look at yourself in the other side of the spoon, the concave side. If you are holding the spoon at a reasonable distance from your face, you should see a reduced, inverted image of you and your surroundings. If you place your finger close enough to the spoon, you will notice that the image inverts, and you form an upright, enlarged, virtual image of your finger. A concave mirror like this will also produce a real image, one that you can focus on a piece of paper or film.

Any mirrored or metallic surface in your house that bulges toward you at the center (like a kitchen faucet) acts as a convex mirror. See what other funhouse-type mirrors you can find in your house.

lenses. The two types of lenses are shown in Figure 10.12.

A single converging lens will cause parallel light rays to meet at a point at a distance beyond its surface, called the *focal length.* Picture a beam of parallel rays of light coming from a distant source along the axis of a converging lens, as shown in Figure 10.13. When the ray of light strikes the mirror on its axis, it slows (owing to the different index of refraction in the glass) but passes through the glass. Off axis, the rays of light strike the glass at an angle, and as they enter the glass bend toward

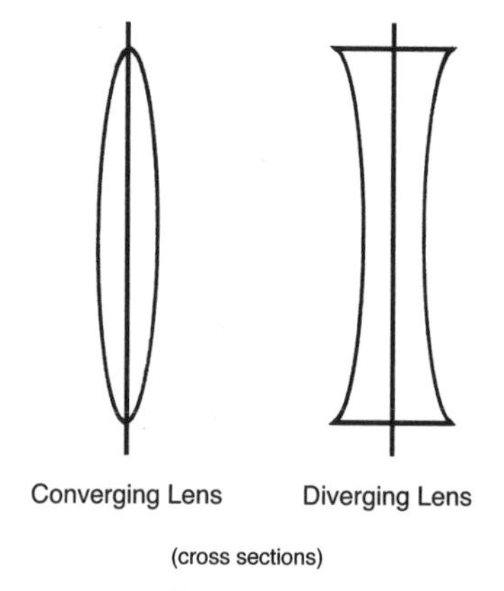

Figure 10.12

PHYSICS IN THE REAL WORLD

Human eyes—made to bend light coming from air into the eye—don't work well underwater; fish eyes—made to bend light coming from water into the eye—don't work well in air. This occurs because the amount of bending depends on the difference in the index of refraction between the two substances. Mermaids would have to have small water-filled goggles over their eyes to see normally outside the water in much the same way that humans need small air-filled goggles to see well underwater.

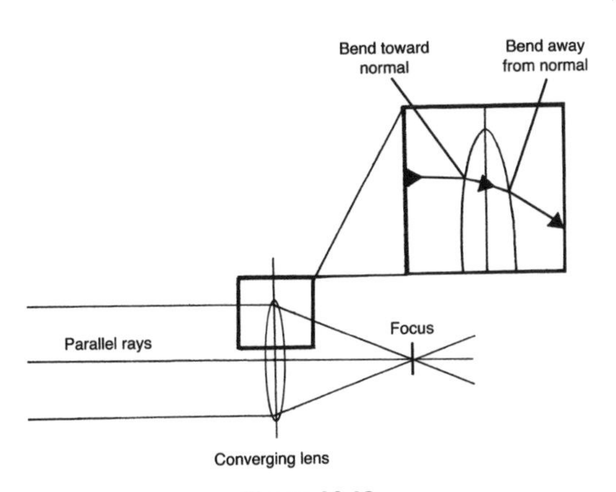

Figure 10.13
Refraction in a Convex Lens

the normal (see Figure 10.13). As the ray exits the glass, it bends again, this time away from the normal to the surface on the other side.

The result is that the incoming ray of light is bent toward the axis to which it arrived parallel. All incoming rays that strike the glass will be brought to a point at the focus of the lens.

Forming an Image

We can see what happens when we try to form an image of an extended object as opposed to a point. Figure 10.14 shows where a converging lens will form an image of an object placed beyond the focal length and within the focal length.

Notice that the real image formed of an object beyond the focal length is inverted. Rays that come from the object parallel to the axis of the lens are bent through the focus of the lens.

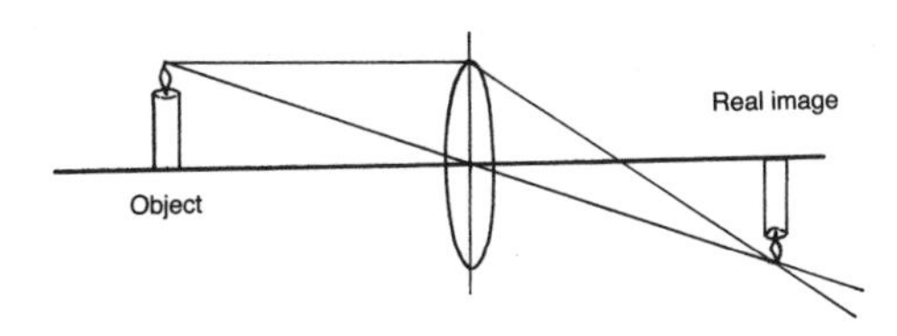

Figure 10.14
Forming a Real Image with a Convex Lens

Rays that pass through the center of the mirror continue in a direction unchanged (assuming that the mirror is thin), and rays that pass through the focus of the lens emerge parallel to the axis of the mirror on the other side.

The drawings that you can generate are called *ray diagrams*. Using these rules, you can use ray diagrams to determine the location of a real or a virtual image formed by a mirror or lens. Figure 10.15 shows examples of ray diagrams and the images formed. The location where rays cross is where a real image is formed. If the rays do not cross (meaning that they diverge or move apart), then the lens will form a virtual image at the location from where the diverging rays appear to come.

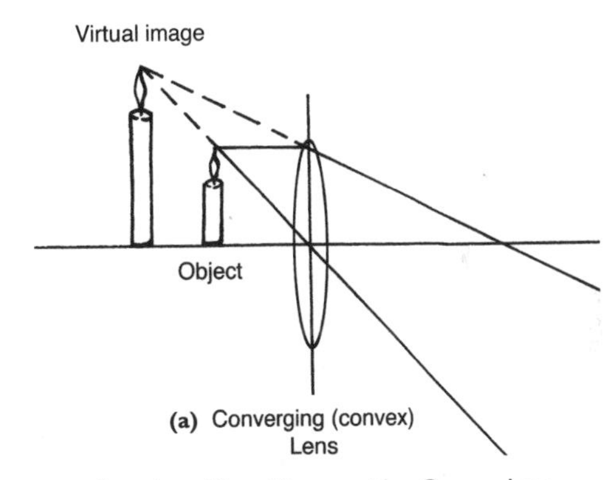

Forming a Virtual Image with a Convex Lens

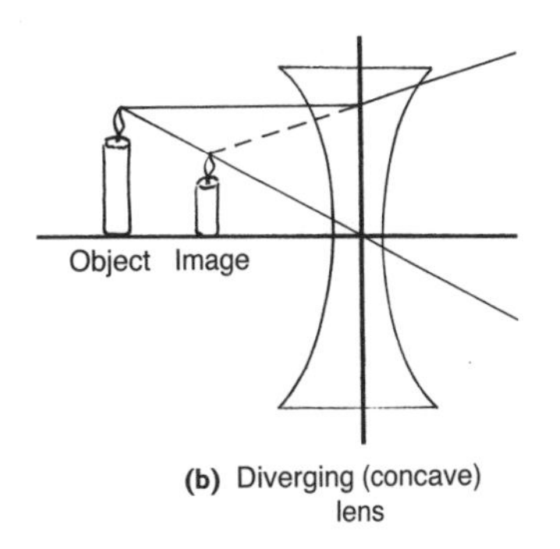

Forming a Virtual Image with a Concave Lens

Figure 10.15

PHYSICS IN THE REAL WORLD

For a converging lens, an object placed anywhere within the focal length cannot form a real image, and one sees a virtual image on the same side of the lens as the object. In this case, the lens is acting as a magnifying glass.

There is also a mathematical way to determine the location and size of an image. The relation giving the distance is called the *lens formula*:

$$\frac{1}{p} + \frac{1}{q} = \frac{1}{f}$$

where p is the distance from the lens to the object, q is the distance from the lens to the image, and f is the focal length of the lens. The size (or height) of the image is given by

$$h_i/h_o = q/p$$

where h_i is the height of the image and h_o is the height of the object.

For example, suppose a candle is 30 in. from a converging lens whose focal length is 10 in. Where will the image of the candle be formed? If the candle is 2 in. tall, what will be the height of the image of the candle formed? Using the lens formula, we obtain

$$\frac{1}{30} + \frac{1}{q} = \frac{1}{10}, \quad \text{or} \quad \frac{1}{q} = \frac{1}{10} - \frac{1}{30} = \frac{1}{15}$$

Inverting both sides of the equation, we obtain the result

$$q = 15 \text{ in.}$$

The formula determining image size gives

$$h_1 = 2 \times 15/30 = 1 \text{ in.}$$

Thus the image of the candle will be located 15 in. beyond the lens and will be half as tall as the object (and inverted).

If you replace the converging lens used in the preceding problem with a diverging lens, you will not form a real image. When parallel light strikes a *diverging lens*, the refracted rays spread apart from one another, and all of them *appear* to come from a point on the near side of the lens (Figure 10.15). This point is a called the *virtual focus*. The distance of this point from the lens is called the focal length, just as for a converging lens. Diverging lenses always form virtual, upright images that are smaller than the actual object. If you wear eyeglasses, hold them up and look through them at the scene in front of you. If you see a reduced version of the scene, then your glasses are functioning as diverging lenses.

OPTICAL INSTRUMENTS

Cameras and the Human Eye

In Exploration 10.1, you put together what is basically a *camera*, except that it had no provision for keeping out unwanted light and no film for making a permanent record of the image. Also, all but the simplest photographic cameras have a combination of lenses rather than a single lens to improve sharpness and other characteristics of the image.

Most cameras also have a provision for varying the size of the lens opening and to regulate the brightness of the image when pictures are taken under various lighting conditions. Until recently, most cameras used photographic film, which contains chemicals sensitive to light. This film was exposed when the camera shutter was opened, and the film was then taken to a store for processing.

A CCD or *charge-coupled device* has replaced the film in many cameras. This is basically a grid of silicon that is sensitive to photons that strike it. In the case of such a *digital camera,* the image is instantaneously available and is generally stored to a removable memory device. Interestingly, the optics described in this chapter have not changed for digital cameras. One still pays a premium for excellent optics in a camera, and one can still buy a cheap digital camera that has lower-quality optics.

The human eye is optically very similar to a camera. In fact, it most closely resembles a digital video camera with a CCD detector and a hard drive to store visual images, since the images it detects are constantly changing. The eyeball contains a lens-shaped organ whose focal length can be shortened by sphincter muscles capable of changing its shape. A watery fluid in front of the lens and a jelly behind it also refract or bend the incoming light, producing an image on the *retina,* the photosensitive rear surface of the eyeball.

Although the image on the retina (a real, inverted, and reduced image) is upside down, our brains have learned to interpret this image right side up. The retina contains millions of delicate nerve endings whose detections of light and color are carried to the brain for interpretation. The *iris* is a

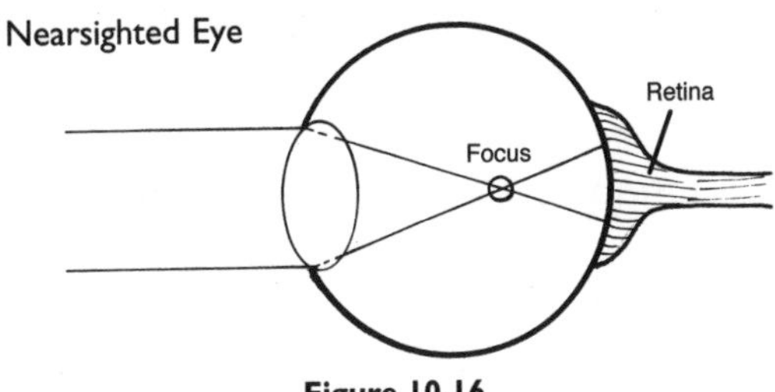

Figure 10.16

variable-size diaphragm controlling the amount of light entering the eye.

Eyeglasses and contact lenses are surely used in greater numbers than any other optical instrument. In a normal eye, the muscles that change the shape of the lens are able to compress it enough so that objects as close as about 10 in. can be seen focused on the retina.

In a *nearsighted* eye, the distance from the lens to the retina is too great. The focused rays cross (forming a real image) short of the retina instead of on it (Figure 10.16).

Nearsightedness can be remedied by placing a diverging lens of the proper focal length in front of the eye, as shown in Figure 10.17. In a *farsighted* eye, unfocused rays strike the retina before they have had the chance to cross—thus forming a real image. In this case, a converging lens of the appropriate focal length can correct the problem.

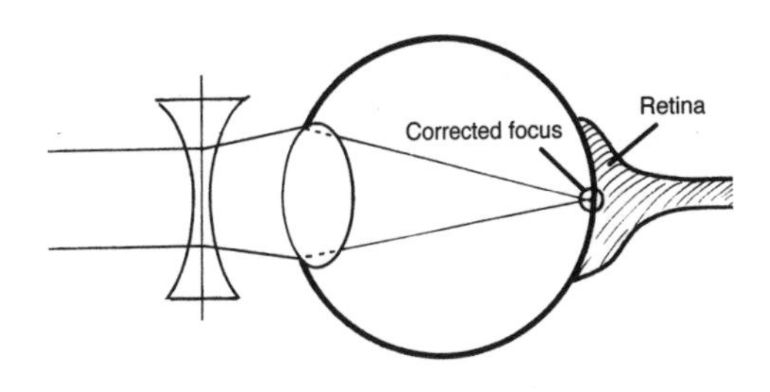

Figure 10.17

In the condition called *astigmatism,* the front surface of the eyeball is not curved equally in all directions like a sphere (recall the spoon in Exploration 10.7), so images are distorted. If you wear eyeglasses, hold them up in front of you and rotate them. If the image that you see distorts as you rotate the glasses, then you have astigmatism. A lens that is curved in one direction only (a cylindrical lens) remedies this condition. If you have both astigmatism and near- or farsightedness, the two kinds of lenses needed for correction can be combined into a single piece of glass.

Microscopes and Telescopes

A compound microscope is a combination of two converging lens systems—an *objective* of very short focal length and an *ocular* or *eyepiece* of moderately short focal length. This arrangement is shown in Figure 10.18 with each system of lenses represented as a single lens for simplicity.

The object to be examined is brought up close to the objective lens system, forming a real image of it inside the tube. The ocular, used as a simple magnifier, then produces an enlarged, reversed virtual image from this real image. Magnifications on the order of a few hundred to a few thousand are available with such a system.

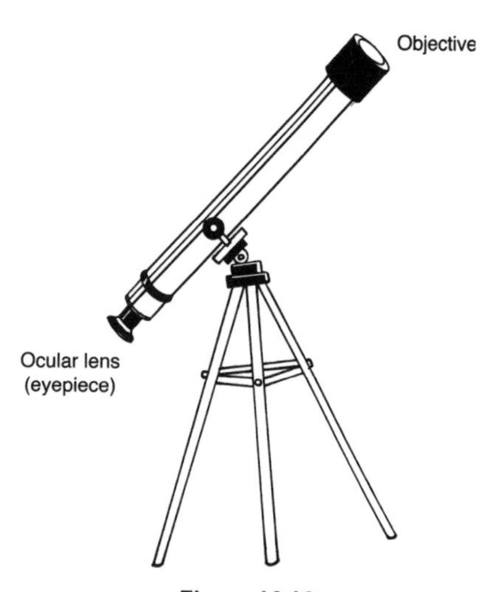

Figure 10.18

A *refracting telescope,* like a microscope, consists of an objective lens system that forms a real image and an ocular or eyepiece for magnifying the image. In the case of a refracting (lens-based) telescope, the objective tends to have a very long focal length. For astronomers, light-gathering power is far more important than magnification. Stars are so distant that magnification is not a critical issue (stars tend to look like points of light regardless of magnification). Also, because of the turbulence in Earth's atmosphere, magnifications of more than a few hundred are seldom used. The light-gathering power of a telescope determines how faint an object can be detected. Light-gathering power depends on the area of the objective lens or mirror and is the primary reason for making large-diameter telescopes.

Most research telescopes are not refractors (employing systems of lenses) but *reflectors,* using a system of mirrors to gather and focus light. The reason for using reflectors is that when astronomers started to make increasingly larger lenses, they discovered that it was difficult to produce large volumes of glass without imperfections, and gravity tended to distort large glass lenses, reducing their performance.

Reflecting telescopes employ a large, curved, primary mirror to collect light and focus it, generally with the help of a secondary mirror. It is far easier to make a flawless surface than a flawless volume. The world's great optical telescopes (e.g., the Keck telescope, the Hubble space telescope) are all reflectors.

PROBLEMS

10.1 What is the speed of light in olive oil? In a diamond?

10.2 Assume that two mountains are separated by 50 mi. How long will it take for the light "released" when a lantern is uncovered on one mountain to reach the other one?

10.3 What luminous intensity (in candles) is required to produce 10 ft-c of illumination 30 ft below a street lamp?

10.4 Draw ray diagrams that explain why images form the way they do when you look into the concave and convex faces of a reflective metal spoon.

10.5 Draw a ray diagram that shows why a spear fisher would perceive a fish to be closer to the surface of a body of water than it actually is.

10.6 You place an object 5 in. from a converging lens with a focal length of 10 in. What type of image will it form? Will the image be larger or smaller than the object?

10.7 Is the main rearview mirror in your car a plane or a curved mirror? How can you tell? How about the driver's side rearview mirror?

10.8 Can you think of any reasons, other than the ones given in the text, why research telescopes employ mirrors instead of lenses?

COLOR AND THE WAVE NATURE OF LIGHT

KEY TERMS

dispersion, spectrum, spectroscope, spectrograph, continuous spectrum, absorption line spectrum, emission line spectrum, redshift, blueshift, electromagnetic spectrum, interference, diffraction grating

PHYSICS IN THE REAL WORLD

Sunlight high in the atmosphere refracts as it passes into tiny water droplets, reflects off the back surface of each droplet, and then refracts again as it emerges from the front of the droplet. When this bending and reflection happens in a large number of droplets in the sky, we see the effect as a rainbow. Why does the bending of light cause a rainbow?

When light bends as it enters and exits the water droplet, the amount of its bending depends on the wavelength of the light. The wavelength of visible light determines its color. Red light (long wavelength) is bent the least, yellow light and green light are bent more, and violet light (with the shortest wavelength) is bent the most. Like light traveling through a glass prism, sunlight (containing all the colors of the rainbow) passing through a water droplet is separated into its component colors.

THE SPECTRUM

If you hold a triangular glass prism in the path of a beam of sunlight, you will—as you may know—create a rainbow of colors. The apparently "white" sunlight actually contains the entire spectrum of colors: red, orange, yellow, green, blue, indigo, and violet. The way students usually remember the sequence of colors from longest wavelengths to shortest wavelengths is with the name Roy G. Biv, which consists of the first letter in the name of each color. Light refracts (bends) as it enters and leaves the glass, much as light does as it traverses rain droplets.

Sir Isaac Newton first established that these colors are all present in the "white" light of the Sun and that, in passing through the prism, each color is refracted by a slightly different amount; thus, the colors spread out. This spreading out of the light is called *dispersion*. The sequence of colors, or any spreading of light by wavelength, is called a *spectrum*. Once the colors are dispersed, the human eye—which blends all the colors together to make white light—can see the individual colors.

Because different wavelengths are refracted by slightly different amounts, some separation of color occurs whenever light passes through any transparent material, and accidental rainbows are a familiar sight in the atriums of buildings with many panes of beveled glass.

EXPLORATION 11.1

You can test Sir Isaac Newton's idea about the nature of sunlight if you have two prisms. Prisms are often available in novelty or hobby shops. Place the two prisms in sunlight, reversing the position of the second prism. The first prism should generate a rainbow of color (test this with a piece of paper); the second prism should combine the spectrum back into a colorless path of white light.

You also can confirm that monochromatic (single-color) light is not further broken up into more colors. Cut a small hole in a card. Allow only one color emerging from the first prism to enter the hole and strike the second prism. You will find that the light is bent (refracted) in the second prism, but its color does not change.

The most familiar example of the dispersion of white light can be glimpsed after a rainstorm. In order to see a rainbow, you must have the Sun at your back so that rays can be refracted on entering droplets of water, reflected from the back surface, and then refracted again as they emerge.

Colors and Pigments

What is color? In terms of visible light, color is simply electromagnetic energy of a specific wavelength. Within the visible part of the spectrum (between 400 and 700 nm, or 4000 and 7000 Å), the color perceived by the eye is dependent on the wavelength of the light. Thus, objects that give off (or reflect) red light look red. Objects that give off (or reflect) indigo light look indigo.

We can easily convert between the wavelength and frequency of light by noting that the relationship among wavelength, frequency, and velocity for any wave is given by

$$v = \lambda f$$

where v is velocity, λ is wavelength, and f is frequency.

Because the velocity of light, c, *is* constant in a vacuum, this relationship becomes

$$c = \lambda f$$

Also, because the speed of light (c) is a constant, wavelength and frequency are inversely proportional; that is, larger (longer) wavelengths are associated with smaller (lower) frequencies, and vice versa.

When we say that the spectrum is made up of the colors red, orange, yellow, green, blue, indigo, and violet, these divisions are for convenience. The visible spectrum actually consists of an infinite gradation of color from red to violet.

PHYSICS IN THE REAL WORLD

In truth, nature is the ultimate crayon box, with the colors varying infinitesimally from red through orange yellow to violet. Computers approximate this infinity of colors with varying degrees of accuracy. Older computer graphics cards produced 256 shades of color (called 8-bit color). Current graphics cards are 16- or 24-bit, offering 65,536 or more than 16 million colors, respectively.

Pigments

The results of mixing paints are altogether different from the results of mixing light. For example, we saw previously that mixing together all of the colors of light in the spectrum produce white light. If you remember art class, however, mixing together all the colors in your paint box certainly does not produce white paint; in fact, it produces a brownish-gray mess.

In terms of light—and pigment, or paint—virtually any color can be produced by mixing proper proportions of the three *primary colors*: red, green, and blue-violet. In fact, the colors on your television screen are produced by clusters of these three colors covering the screen, which you will see if you look closely at your screen. If you project these three colors on a screen and let them partially overlap, you will find that the region of overlap of all three colors is white.

Paints or pigments function by *absorbing* (removing) some wavelengths from white light striking the surface and reflecting the other wavelengths. Therefore, an object that looks red is one that absorbs all wavelengths except red, which it reflects. Its surface has a red color when illuminated by white light.

Thus, when two colors of paint are mixed, the result is a color that is the combination of the two colors of that paint. Imagine that you mix together some yellow and some blue paint. We all know that the result is green paint. This is because the yellow paint absorbs blue and violet wavelengths, and the blue paint absorbs red and yellow wavelengths, meaning that only green wavelengths are not absorbed but are reflected. Mixing together all colors in the paint set will mean that the resulting sludge will effectively absorb all colors and start to look gray or black (depending on how effectively the colors absorb across the spectrum).

EXPLORATION 11.2

What would a red shirt look like under yellow light? Many streetlights in cities are "low-pressure sodium" lights, which are a combination of filament and gas. Their light is in the yellow part of the spectrum, and astronomers like these lights in particular because they can be more easily filtered than many other types of lights.

If you have such a streetlight near you, put on a red shirt and go outside. Your shirt should look black. Why is this? The red pigment in your shirt absorbs all colors except red, which it reflects. If the only light striking your shirt is yellow, then your shirt will reflect none of the incident light and will appear black.

Different Types of Spectra

Any instrument used to view a spectrum of light from a source is called a *spectroscope*. In a classic spectroscope, light is passed through a thin slit, through a prism, and then magnified to be viewed at an eyepiece. If the goal is to record the spectrum for analysis, then in place of an eyepiece either photographic film or a CCD (digital) camera is used to record the pattern. If the spectrum can be recorded, the instrument is called a *spectrograph*.

Different types of light sources produce different spectra, as can be seen with even a simple spectroscope. Objects like a lightbulb filament, the Sun, or the heating element from a stove will produce a spectrum that appears to

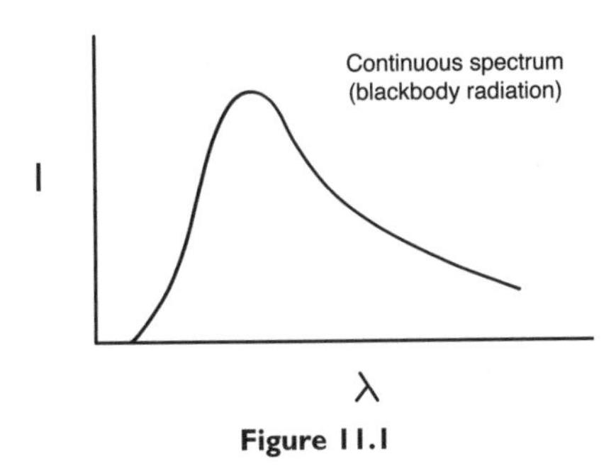

Figure 11.1

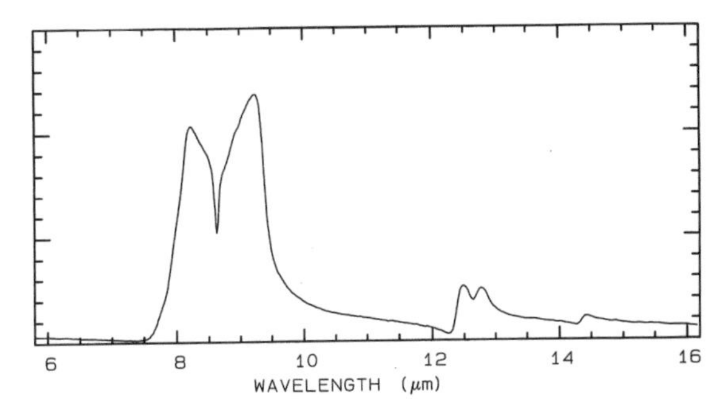

Figure 11.3

contain all the colors of the rainbow, which is also called a *continuous spectrum*. Figure 11.1 shows a continuous spectrum as a plot of intensity as a function of wavelength.

Many spectra that appear to be continuous actually contain gaps; that is, they are within the continuous spectrum wavelengths that are not represented, and appear as dark bands. The spectrum of the Sun (examined in detail) contains many dark bands called *absorption lines*, which produces an *absorption line spectrum* as shown in Figure 11.2.

Other spectra consist of only bright lines; this type of spectrum is called an *emission line spectrum*, and is shown in Figure 11.3.

Each of these types of spectra arises from different physical situations. A continuous spectrum arises from any hot object (like the Sun or the heating element on a stove). The ideal version of such an object is sometimes called a *blackbody*. A blackbody is simply a perfect absorber and emitter of radiation. The wavelength at which the intensity of a blackbody peaks determines its temperature. Objects whose peak is closer to the red end of the spectrum are cooler, and those whose peak is closer to the blue end of the spectrum are warmer.

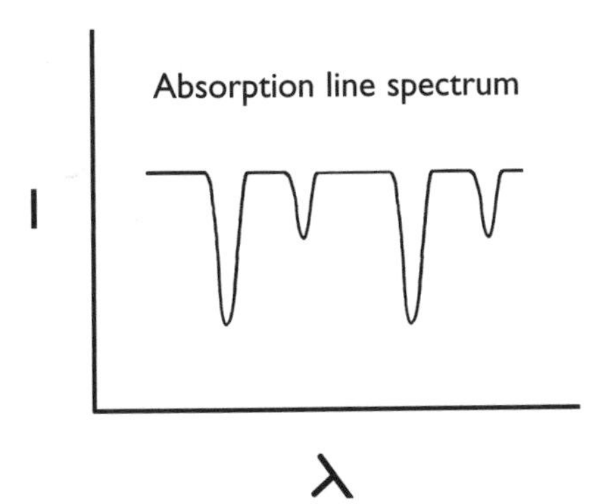

Figure 11.2

An absorption line spectrum arises when a hot object (like the Sun) is behind a cooler object (like the outer layers of the Sun's atmosphere). In this case, the intervening gas absorbs light at particular wavelengths, so these wavelengths get subtracted from the continuous spectrum. The exact colors (wavelengths) absorbed provide evidence about the atoms and molecules that are present in the outer layers of the Sun.

An *emission line spectrum* is produced when the gaseous form of an element is excited by some energy source. As the element returns to its nonexcited or ground state, it emits characteristic, signature wavelengths (and therefore colors) of light. The pattern of lines produced relates to the jumping of electrons from higher to lower energy levels within the atoms. Different jumps release different amounts of

PHYSICS IN THE REAL WORLD

The element helium (He) was first discovered (and named for) the Sun (Helios). It was thought that this element was unique to the Sun until it was discovered on Earth. Helium (being a noble or inert gas) is nonreactive and does not combine with oxygen in the atmosphere in the dramatic way that hydrogen can. The composition of many distant sources is also determined by spectroscopy. Radio telescopes have been used, for example, to detect the presence of complex molecules in distant clouds of gas from which stars are forming. Earth-bound physicists and chemists know the lines that identify certain compounds, and detecting these same spectral lines from a source in space tells us that those elements are located in those regions as well.

energy. We will discuss this phenomenon in more depth in Chapter 15. The pattern of emission lines produced is different for each element in the periodic table and each molecule that these elements can form. Once we know the characteristic pattern for a given element, then we can use that knowledge to identify elements present in any object from which we can gather light. This, in a nutshell, is the power of astronomy. We can determine the elements and molecules present in sources that are far too distant to ever visit simply from the light that reaches us from that source.

An even more detailed analysis of the spectrum from a distant source (not just the presence of certain spectral emission lines but their relative brightness and detailed shapes) can tell us even more about the conditions in a distant source, including its temperature and pressure, as well as its motions relative to us.

DOPPLER EFFECT FOR LIGHT

In Chapter 9, we discussed the Doppler effect, or the change in the frequency of sound as a result of the motion of the source. Very much the same effect is observed for sources of light. When an object is in motion relative to Earth, the light that it emits is *redshifted* (to longer wavelengths) if it is moving away from us and *blueshifted* (to shorter wavelengths) if it is moving toward us. The shifts in wavelength are small but measurable.

EXPLORATION 11.3

The rest wavelength for a bright red line in the emission line spectrum of the hydrogen atom is 656.3 nm. If an astronomer observes this line in a galaxy moving away from Earth at 100 km/s, what will be the change in wavelength, and will it appear as a shorter or longer wavelength?

First, if the source is moving away, the light will be redshifted, making the detected wavelength longer. Thus, we use the relationship

$$\Delta\lambda/\lambda = v/c$$

and multiply both sides by the wavelength to get

$$\Delta\lambda = \lambda v/c$$

Being careful to use the same units for the speed of light as our velocity (km/s), we obtain

$$\Delta\lambda = (656.3 \text{ nm}) (100 \text{ km/s})/(3 \times 10^5 \text{ km/s})$$

or

$$\Delta\lambda = 0.2 \text{ nm}$$

In fact, the change in wavelength is related directly to the speed of the source and the speed of light. The change in wavelength ($\Delta\lambda$) is related to the rest wavelength (λ) in the following way:

$$\Delta\lambda/\lambda = v/c$$

where v is the speed of the source relative to the observer and c is the speed of light (3×10^8 m/s).

THE ELECTROMAGNETIC SPECTRUM

The human eye is sensitive to a rather limited range of wavelengths or frequencies—namely, 400 nm to 700 nm. It is insensitive outside of that range, and many physical processes in the world give rise to wavelengths that are outside that range. Wavelengths that are longer than 700 nm are referred to as *infrared,* and those that are shorter than 400 nm are referred to as *ultraviolet.*

X-rays have wavelengths still shorter than ultraviolet waves. As wavelengths get shorter (and frequencies increase) the energy carried by light gets more and more intense, as we will see in Chapter 15. X-rays, with their high frequencies, are high-energy waves and are able to penetrate most materials, making them useful for imaging bones through skin, for example.

Gamma rays have shorter wavelengths and are more energetic than X-rays. Gamma rays are given off in some of the most energetic events known, including the explosion of atomic weapons and the explosive collapse and death of a star, called a *supernova.*

Having wavelengths that are longer than optical or infrared radiation, radio waves are used to transmit information in television and radio broadcasts, as well as in cordless and cellular phones. Because radio waves are long wavelengths, they emit a low frequency, and typically are thought not to be harmful to humans.

All the types of waves described constitute the *electromagnetic spectrum,* and like all waves, they can be reflected, refracted, and diffracted. All these different types of waves travel through a vacuum with the same speed (c).

INTERFERENCE AND COLOR

If you have ever seen a thin film of oil floating on a pond or puddle in the sunlight, you may have noticed that it is filled with a rainbow of colors. This colorful display, also seen in soap bubbles, is the result of *interference* between light that is reflected off the top surface of the oil (or soap) and light that is refracted at the top surface, reflected off the bottom surface, and then refracted back out of the top. We discussed interference when we first introduced waves (in Chapter 9). Constructive and destructive interference can occur when two waves are in the same space, but are out of *phase,* that is, at different points in their cycle.

In the case of waves on a string, this might mean that as two upright waves pass each other, their amplitudes will add, making for a moment a wave with double the amplitude. Light can interfere in much the same way if its phase is changed. Thus, if two light waves are in phase (maximum lined up with maximum), then they will constructively

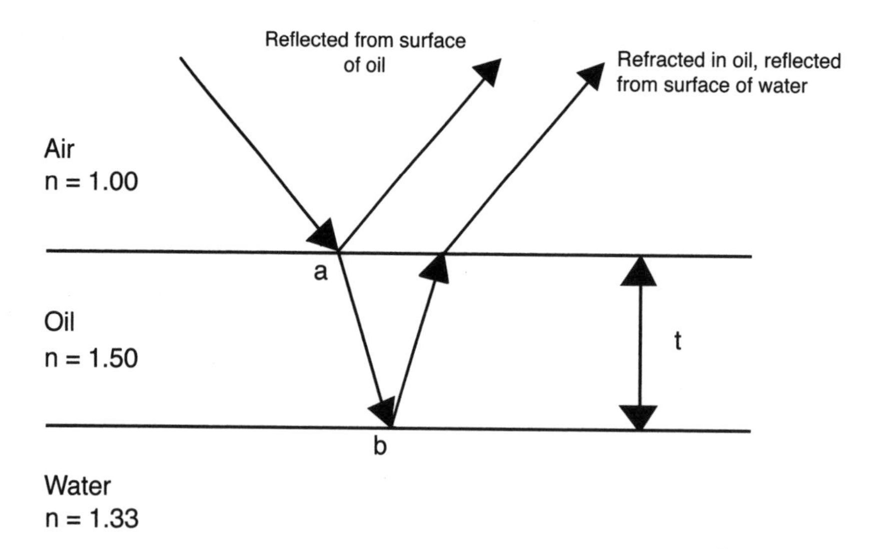

Figure 11.4a

interfere (making a brighter spot). If two light waves are out of phase (maximum lined up with minimum, for example), they will destructively interfere (making a dark spot).

In these examples of interference, light from single source is reflected from two surfaces that are close together. When you look at oil on water, for example, part of the light that comes to you is reflected from the top surface of the oil and part from the bottom. Figure 11.4a shows this process.

If the two reflected rays of a given wavelength emerge from the oil completely out of phase, they will exhibit complete destructive interference and no light of this color will emerge from this point of the film. Whether interference of light will occur at that point on the surface depends on two factors: the wavelength of the light and the thickness of the oil at that specific spot. Because white

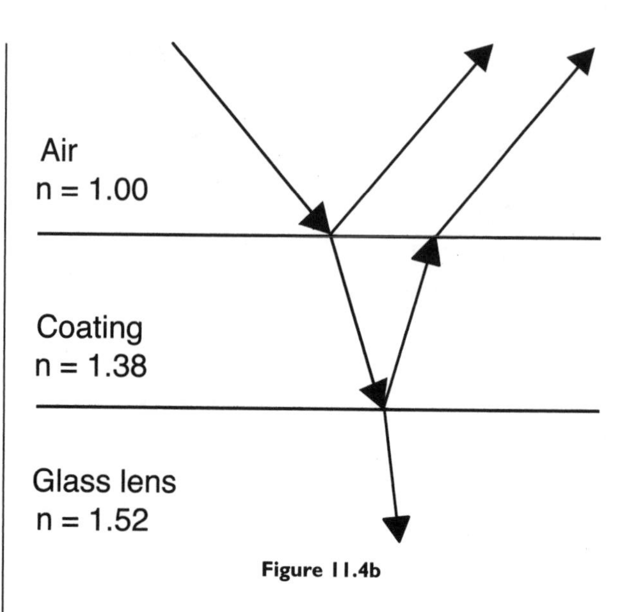

Figure 11.4b

light consists of many different wavelengths (as we have seen), and since the oil floating atop the water is not uniformly thick, you will tend to observe a rainbow of colors, moving in complex patterns across the oil.

PHYSICS IN THE REAL WORLD

A common application of optical interference is the use of nonreflective coatings for glass in camera lenses, for example. In this case, the surface is covered with a coating of just the right thickness to cause destructive interference for most of the light that would ordinarily be reflected from the surface of the glass. Glass that has been coated in this way will sometimes have a violet or purple hue when seen in sunlight (see Figure 11.4b).

DIFFRACTION

Water waves bend when they pass through barriers, and sound waves clearly bend around obstacles like walls and buildings; otherwise, we would hear much less in our daily lives.

If light could be understood completely as rays of light traveling in straight lines, then when rays of light passed through a thin slit and struck a wall, the result should always be a shadow (from the barrier) and a thin patch of light (from the slit). Experiments show, however, that as the slit width narrows, there is a thickness at which the thin patch of light actually starts getting thicker. This spreading of light through a thin slit is known as *diffraction*. In fact, diffraction is the bending of any kind of wave when it passes near the edge of an obstacle or through a small opening. In the case of light, the opening must be small indeed for the diffraction to be noticeable. Also, if the incident light has a single wavelength (like laser light, sometimes called coherent light), the diffraction becomes far more noticeable.

EXPLORATION 11.4

Many people now own laser pointers. These are small sources of laser light that have a specific wavelength—the most inexpensive being in the red part of the electromagnetic spectrum. If you have a laser pointer, you can use it to observe the diffraction of light.

In a darkened room shine the laser pointer at a white wall. You will see a spot of intense red light. Now, slowly raise the edge of a knife into the laser beam so that the edge of the beam strikes the knife. The pattern on the wall will broaden, indicating that the light is diffracting at this sharp edge.

When light is diffracted through a small opening, it produces a series of bright and dark lines when projected onto a wall or screen. If the incoming light is all one wavelength, then the pattern of light and dark will be the same color; however, if the incoming light is white light (sunlight, for example), then each of the alternating patterns of light and dark will be a spectrum, showing the colors from red to violet. This type of pattern will arise from a single slit, a double slit, or a whole series of closely spaced slits. When many slits are placed close to one another, the material is called a *diffraction grating*. Early diffraction gratings were made of closely spaced wires, though the same effect can be accomplished by etching narrow grooves in a plate of glass. If light passes through the grating and is projected onto a screen, it is called a *transmission grating*. If light is reflected off a grooved surface, which produces a similar pattern, then it is called a *reflection grating*. Reflection gratings are more commonly used and, in fact, have almost completely replaced prisms.

EXPLORATION 11.5

If you have a CD handy, hold it up and look at the bottom side of the CD as if it were a mirror. You will see a rather poor reflection of yourself, but also a series of radial rainbow patterns. These patterns are the result of the concentric rings, which act like reflection grating on the surface of the CD.

If you reflect lamplight from the surface of the compact disk onto a wall in a darkened room, you should easily make out a series of circular spectra, varying from blue to red. The innermost circular spectrum will be the most intense, but you will be able to see fainter spectra farther from the center of the reflection.

Surfaces very much like that of a CD are used in spectrometers and spectrographs in industry, except that the alternating patterns of reflection and absorption are not circular.

Optical reflection gratings are too coarse to produce diffraction from very short wavelength radiation. For X-ray radiation, smaller structures must be found; in fact, the atomic spacing in the crystalline structure of certain materials can produce diffraction patterns from X-rays. These diffraction patterns have been used to determine the structure of crystals as well as the structure of other molecules.

Rosalind Franklin used X-ray diffraction to explore the structure of the DNA molecule. The tiny spacings between atoms in this molecule are sufficiently small to diffract the tiny wavelengths associated with X-rays. "Photo 51" is the famous image that was used to first suggest that the DNA molecule had a double-helix structure, like a double-winding staircase. Unknown to Franklin, Watson and Crick were shown this photograph by one of her collaborators, Maurice Wilkins. Watson, Crick, and Wilkins went on to receive the Nobel Prize in physiology or medicine for determining the structure of the DNA molecule. Franklin died at the age of 37 in 1958—five years before they received the prize.

PROBLEMS

11.1 Can you explain why it is possible to see a circular rainbow when you are in an airplane, yet it is impossible to see it from the ground?

11.2 What is the frequency of red light, assuming that its wavelength is 656.3 nm?

11.3 What is the frequency of radio waves, assuming that their wavelength is 21 cm?

11.4 What color would a banana appear under yellow light? Under blue light?

11.5 If you are approaching a stoplight, how fast will you have to be traveling to make a red stoplight appear green? *Hint*: Assume that the actual wavelength is 700 nm (red) and that you have to be going fast enough to shift it to 500 nm (green).

11.6 Why do you think reflection gratings are much more widely used than prisms for spectroscopy?

11.7 What sort of spectrum would you expect to observe from a streetlight and why?

11.8 What sort of spectrum would you expect to see from a molten metal and why?

CHAPTER 12

ELECTRIC CHARGE AND POTENTIAL

KEY TERMS

static electricity, electric current, coulomb, electrical force, conductor, insulator, grounding, induction, electric field

PHYSICS IN THE REAL WORLD

As you walk down the sidewalk on your way to class, your cell phone vibrates. As you take it out of your pocket, you take the earpiece from your headphones to your MP3 player out of your ears and answer the phone. It is your friend calling about dinner tonight. You pull out your PDA and flip to today's schedule and see that you are free for dinner. You say good-bye to your friend and then wait at the corner for the light to change. As you wait a nearly silent automobile hums past, powered by an electric motor. The pedestrian light changes to WALK and you cross the street, heading on your way. Lightning flickers in the distance, so far away that you do not hear thunder, but you check to make sure that you have your umbrella with you in your backpack. It's getting warmer, so you quickly pull off your sweater, feeling a crackle in your hair as you do so. You wonder if your hair is standing on end now.

Electric charge, electric currents, batteries, and electromagnetic signals are so much a part of our daily lives that we barely notice them anymore. Most of the time, we are surrounded by electrical phenomena in both the natural world and the human-made world. We learned about the presence of electricity in the world through experimentation and observation, and we have learned how to harness, control, and store electrical power for our everyday use, as in the tiny batteries that power our watches and electronic equipment.

Long ago, Greek scientists discovered that a material called amber when rubbed with a cloth would attract small pieces of straw. In later centuries other materials were found to have this same characteristic. There was some property of the cloth that apparently was transferred to the amber when the two were rubbed together, and that property enabled the amber (for a brief time) to attract small pieces of straw. The amber had been *electrified.*

In this chapter and the next, we explore the branch of physics called *electrostatics,* which studies the nature of charge that is not moving, and set the stage for a consideration of moving charge that is needed to understand the electrical circuits in all the electronic devices that we use every day.

STATIC ELECTRICITY AND ELECTRIC CURRENT

All of us have experienced the phenomenon of static electricity. If you run a comb through your hair in dry weather (or remove a sweater on a dry day), you will sometimes hear a crackling noise. When you have been sitting in a car with cloth seats, you might get a slight electric shock when you touch the metal of your car door as you slide across the seat to get out. *Static electricity* is simply the presence of electric charge (positive or negative) on the surface of a material.

Any two substances that are rubbed together can potentially become charged. If the substances are initially electrically neutral (as most substances are), a transfer of charge will give one substance a net positive charge and one of them a net negative charge. *Electric current* results from the movement of electric charges.

Experiments in the eighteenth century showed that a stick of hard rubber could be electrified by rubbing it with a piece of fur, and that, likewise, a glass rod or tube could be electrified by rubbing it with a silk scarf. Like amber, both of these rods will attract small pieces of paper (akin to the straw that the Greeks used); however, the hard rubber and the glass rod are not identically electrified. Further experimentation showed that two electrified pieces of hard rubber will repel each other, but an electrified piece of hard rubber rod and an electrified piece of glass will attract each other.

What had been discovered by these experiments was, in fact, a new force. Like gravity, this electrical force was exerted between two objects; unlike gravity (at least as we currently understand it), the electrical force can be either attractive or repulsive. Objects with the same charge (both positive or both negative) repel one another, and objects with opposite charge (one positive and one negative) attract one another. Benjamin Franklin first gave the names *positive* and *negative* to the two opposite sorts of charge. This was an arbitrary decision, because the only absolute is that there are two different types of charge. Franklin called the charge found on the glass rod positive (+) and the charge on the hard rubber, negative (−).

The SI unit of charge is the *coulomb* (C), and a charge of one coulomb is equal to the charge of 6.25×10^{18} electrons. Taking the inverse of this number reveals that the charge of a single electron is tiny: 1.6×10^{-19} C.

ATOMS AND ELECTRIC CHARGE

The nature of the electrical force is fundamental to an understanding of chemistry, molecular structure, and the structure of the atom. Experiments to probe the strength of the electrical force have found that, over short distances, it is far stronger than the force of gravity and it is intimately related to the generation of electromagnetic waves.

Nineteenth- and twentieth-century experiments showed that there are particles smaller than the "indivisible" atom of the Greek philosophers. Our current understanding of the atom, in fact, is that it consists of three types of particles, two of which can be broken down into yet more fundamental ones.

All atoms consist of a nucleus that contains, at the very least, a proton. Atomic nuclei generally contain both protons and neutrons. Protons have a positive charge and neutrons have no charge; thus, in general, the nucleus of an atom has a positive charge. Negatively charged electrons surround the nucleus of an atom, and under normal conditions, atoms

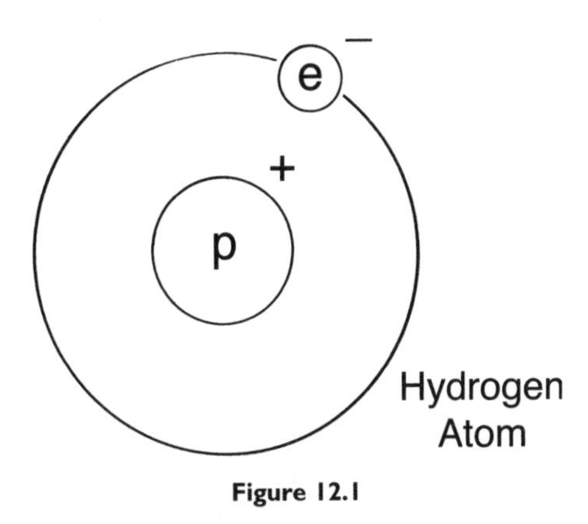

Figure 12.1

are neutral, with the positive charge of the nucleus (from the protons) being equal to the negative charge of the electrons. Such atoms are *electrically neutral.* If an atom has an excess or a deficit of electrons, it is referred to as an *ion.* A model for the hydrogen atom is shown in Figure 12.1.

The electrons that surround the nucleus are indistinguishable and are held to the nucleus by an electrical force that decreases with distance squared, much like the gravitational force.

This model explains how materials can end up with net positive or negative charge. The outermost electrons in many materials are relatively loosely bound (since the electrostatic force drops off with distance squared), and they can be easily knocked loose. When some materials are rubbed together, one material may be better able to hold onto extra electrons than the other. This material accepts extra electrons, and the other material loses some electrons, giving a net negative and positive charge to the two substances.

The nuclei of the atoms in the glass, rubber, silk, and fur are unaffected; only their outer layers (where the electrons reside) are in contact.

COULOMB'S LAW

Coulomb's law describes the *electrical force* between two charged particles separated by a distance r (see Figure 12.2) as varying inversely with the square of the distance, exactly like the gravitational force that we have discussed. Charles Coulomb (1736–1806) discovered this relationship, which can be stated generally as

$$F_C \sim q_1 q_2 / r^2$$

This proportionality becomes exact if we write it with the constant

$$F_C = k q_1 q_2 / r^2$$

where F_C is the electrical force (called the Coulomb force), k is the Coulomb constant, q_1 and q_2 are the charges of the two objects exerting a Coulomb force on each other, and r is the separation between the two objects. In SI units,

$$k = 8.9875 \times 10^9 \frac{\text{N}\cdot\text{m}^2}{\text{C}^2}.$$

You might well wonder why the atom does not simply collapse, because of the mutual attraction of the electron and the proton, and why nuclei of atoms don't fly apart, filled as

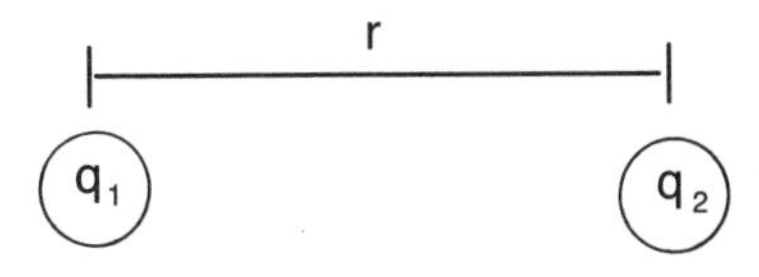

Figure 12.2

EXPLORATION 12.1

What is the strength of the Coulomb force between the two components—an electron and a proton—of a hydrogen atom? Is the force attractive or repulsive? The mean distance between the proton and the electron is tiny, about 5.3×10^{-11} m.

A proton has a charge of 1.6×10^{-19} C, and an electron has the exact opposite charge (-1.6×10^{-19} C). Because the charges are opposite, the force is attractive.

To calculate the force, we simply use

$$F_C = kq_1q_2/r^2$$

Thus,

$$F_C = 8.9875 \times 10^9 \text{ N}\#\text{m}^2/\text{C}^2(1.6 \times 10^{-19} \text{ C}) \times (-1.6 \times 10^{-19} \text{ C})/(5.3 \times 10^{-11} \text{ m})^2$$
$$F_C = 8.2 \times 10^{-8} \text{ N}$$

they are with positively charged nuclei. Don't they all repel one another?

It turns out that there are other forces at work holding the nucleus of the atom together, and that electrons can be thought of as having angular momentum that keeps them in their orbits. We will discuss these finer points of the structure of the atom in Chapter 15.

CONDUCTORS AND INSULATORS

Materials that easily allow charged particles to pass through them are called *conductors*. Metals are good conductors of electricity (and heat). Conversely, materials that do not easily allow charged particles to pass through them, such as the glass and hard rubber in our example of charge transfer, are called *insulators*. If a charged body is to retain its charge, it must be supported by a good insulator. If it is not, all the extra charge that builds up on it will simply flow out, like water out of a drain. Insulators act like drain plugs, holding in (or out) excess charge.

PHYSICS IN THE REAL WORLD

Metal wires are good conductors, and plastic is a good insulator. For this reason, the cords that supply electrical current to the incandescent bulbs in your lamps are made of plastic-wrapped wire. Thus, electric current can travel freely through the wire but not into your hand (and through you to ground) if you happen to hold on to a plugged-in wire.

CHARGING BY FRICTION, CONTACT, AND INDUCTION

We have described how a material can be charged by *friction*. Simply rubbing two materials together (if one of them is a good insulator) can transfer charge from one to the other. You may have on occasion transferred charge from your hair to a balloon by rubbing the balloon across your hair. The balloon ends up with a net negative charge and your hair, with a net positive charge. The balloon can then be made to stick to the wall by induction, and the strands of your hair, all of which are now positively charged, will repel one another, causing you to look a bit odd (see Figure 12.3). On other occasions you may also have transferred charge to your body by scuffing around in socks on a carpet.

Simple contact can also be used to transfer charge. Imagine that you have built up a

Figure 12.3

negative charge in a hard rubber rod. If you touch this rod to an *insulated* metal sphere, the excess charge will flow from the rod to the sphere, giving the sphere a net negative charge (assuming that it was neutral before) by *conduction*. The insulation of the sphere simply assures that the charge will not flow out of it. If the sphere were not insulated, it would be connected to Earth (also called *ground*) by a wire and would be *grounded*.

Grounding is the process of connecting an object to a very large conducting body (like Earth) by means of a conductor (usually a piece of wire). Earth then effectively acts as a well that can accept or provide electrons. Grounding can prevent the buildup of excess charge in materials that are sensitive to the presence of electric charge, like electronics.

Local charges can be induced in a neutral body simply by bringing it near a charged one. *Induction* is the process of bringing a charged object near a neutral one so that the neutral object becomes polarized. In Figure 12.4, a negatively charged rod has been brought close to a neutral, spherical conductor that is insulated from ground. As the negatively charged rod is brought close to the sphere, a positive charge is induced on the near side, resulting in a negative charge on the far side. This induced charge produces an attractive force between the two objects.

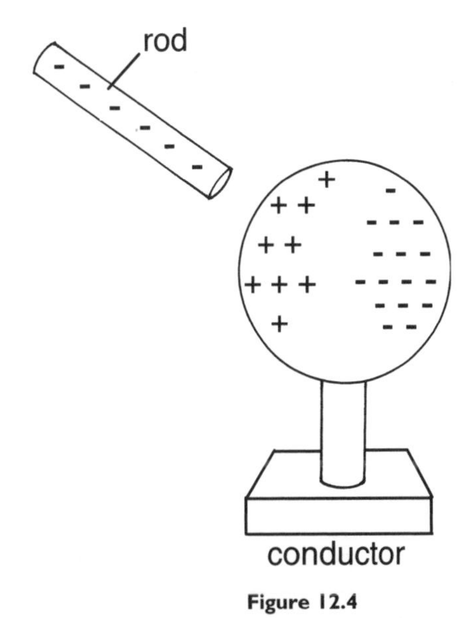

Figure 12.4

EXPLORATION 12.2

If you have a balloon available, inflate it and rub it vigorously across your hair. Rubbing a balloon in your hair will transfer negative charge from your hair (like fur) to the balloon (like a hard rubber rod), leaving the balloon with a net negative charge. Now, place the balloon against the wall. It should stick. Why does it do this?

The balloon begins with a net negative charge. When you bring the balloon close to the wall, it induces a positive charge locally in the wall, as shown in Figure 12.5. The positive and negative charges attract each other, and for a few moments, the attractive force is sufficient to hold the two together. Because neither the balloon nor the wall is a good conductor, both objects can maintain their opposite charges for a short time.

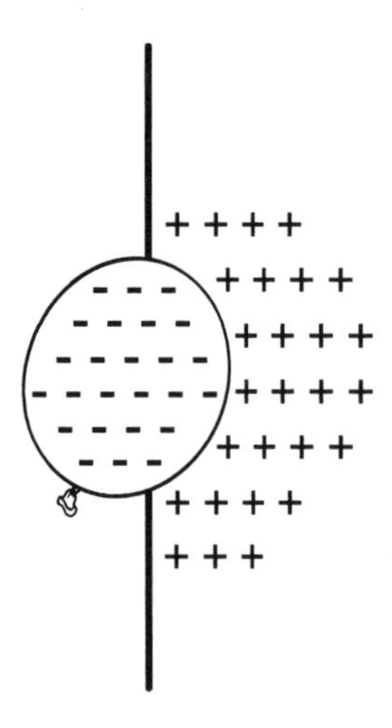

Figure 12.5

If the two objects are allowed to touch, the net negative charge of the rod is partially transferred to the neutral sphere. Now, both objects have a negative charge and repel each other. Attraction changes to repulsion.

Induction also explains the attractive quality of amber for bits of straw. The amber induced an opposite charge in the straw, resulting in an attractive force between the two.

The phenomenon of lightning is thought to result from induction also. Storm clouds become negatively charged on the bottom and positively charged on the top due to convective motions in the clouds. The negative charge on the bottom of the cloud layer induces a positive charge in the ground below it. If the build up of charge is sufficiently great, lightning can occur between oppositely charged parts of a cloud, or between the cloud and the ground below it.

THE ELECTRIC FIELD

We have seen that the electrical force, like gravity, can act between objects that are not physically in contact (see preceding discussion of the atom). In fact, any two charged particles will exert forces on each other; but even a lone charged particle is surrounded by what we call an *electric field* that is a direct result of its net charge. The electric field (like the gravitational field of an object with mass) has both a magnitude and a direction; that is, it is a vector quantity.

The electric field is measured in units of force per unit charge, so that a charge q that experiences a coulomb force F is in an electric field of strength

$$E = F/q$$

where F is the coulomb force and q is the charge of the particle. The SI unit for electric field strength is simply N/C.

Because the electric field represents a vector quantity, we can draw *field lines* that represent the motion that a positively charged particle would take if it were nearby. For this reason, positively charged particles have field lines emanating from them, and negatively charged particles have field lines that point toward them.

Electric field lines are always drawn as though they originate at positive charges and terminate at negative charges. Figure 12.6 shows examples of field lines around a positive charge, a positive and a negative charge, and two oppositely charged plates.

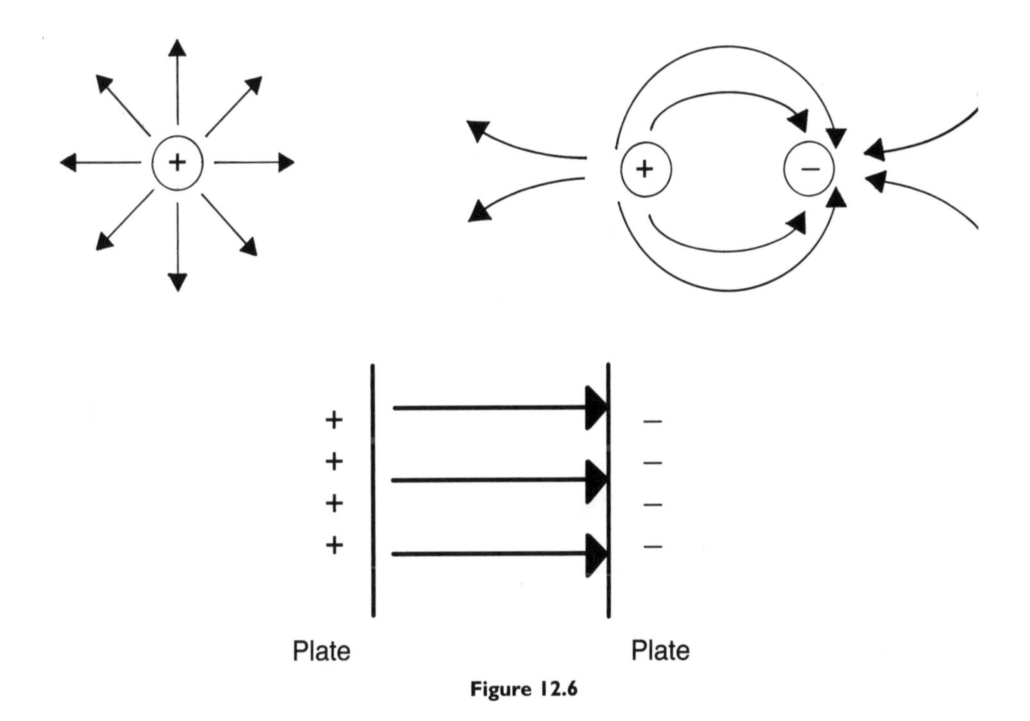

Figure 12.6

ELECTRIC POTENTIAL ENERGY AND POTENTIAL DIFFERENCE

In the same way that physical objects tend naturally to move from a region of high potential energy to a region of low potential energy (e.g., a dropped ball falls down, not up), charged particles move naturally from higher to lower electric potential energy states. To move a negatively charged particle along field lines requires work. Work, you will recall, is defined as a force exerted over a distance. The force required to move a negatively charged particle along electric field lines is stored in the form of *electric potential energy*.

Doing this work results in a change in the potential energy of the particle. If a particle moves between two different states of electric potential energy, we refer to this as the *potential difference* between the two states. The potential difference (also called the *voltage*) is simply the change in potential energy between the two states, divided by the charge:

Potential difference = Change in electric potential energy/Charge

Note that electric potential energy is a scalar quantity, as is potential difference. Because the change in potential energy is measured in joules, and the charge is measured in coulombs, the SI unit of potential difference is simply J/C, where 1 J/C is also called 1 volt (V).

For example, let's calculate how much energy is required to move an electron (with a charge of 1.60×10^{-19} C) through a potential difference of 1 V.

PD = Change in electric potential energy/Charge

1 V = Change in EPE/1.60×10^{-19} C

Therefore,

$$\Delta EPE = 1.60 \times 10^{-19} \text{ C}\cdot\text{V, or } 1.60 \times 10^{-19} \text{ J}$$

This tiny amount of energy is sometimes called an *electron-volt* (eV).

PROBLEMS

12.1 If you were to rub a balloon in your hair and try to stick it to a metal can, would the balloon stick? Why or why not?

12.2 In Exploration 12.1, we calculated the coulomb force between an electron and a proton. Calculate the gravitational force between the same two particles using the universal law of gravitation.

12.3 Name some materials that are good electrical conductors and some that are good electrical insulators.

12.4 Why do you think static electricity can be harmful to electrical equipment?

12.5 How many protons are required to equal a charge of 1 C?

12.6 Given the definition of the word *ion*, what would you say is required to "ionize" an atom?

12.7 Draw the field lines that describe the electric field between two negatively charged particles.

12.8 Draw the field lines that describe the electric field between a positively charged plate and a negatively charged particle.

12.9 How much work is required to move a 2 C charge across a potential difference of 1.5 V?

12.10 What potential difference is present between two plates if 10 J of work is required to move a 0.5 C-charge between them?

CHAPTER 13

ELECTRIC CURRENT AND SIMPLE CIRCUITS

KEY TERMS

electrical potential energy, electric current, electromotive force, direct current, alternating current, resistance, electrical circuit, Ohm's law, power, series, parallel, capacitance

ELECTROMOTIVE FORCE AND ELECTRIC CURRENT

Picture an old-fashioned shower: a bucket of water suspended over your head, perhaps from the branch of a tree, with a rope attached that you pull to release the water in a stream. The water flows downward in this case because of the force of gravity. The water above your head has a type of potential energy that we have discussed in previous chapters called *gravitational* potential energy. The water got that potential energy through the work exerted in pulling the bucket up into the branch above your head.

We can think of *electrical potential energy* in the same way. In the previous chapter we discussed static charge—that is, charge that does not flow; however, if charge is able to flow (because of the presence of a conductor), it will flow from higher to lower potential, in the same way that water flows downhill. In the production of lightning, for example, an *electric current,* or flow of electrons, results when the potential difference between the cloud and the ground gets sufficiently large.

Once potential difference is equalized, because of the flow of electrons, the current will stop flowing. To keep a current flowing there must be a way to artificially (or naturally) maintain a potential difference between two points. In the case of lightning, the potential difference is maintained by the internal dynamics of certain types of clouds. Rubbing a glass rod with silk transfers electrons from the rod to the silk, which creates a potential difference between these two materials. Touching the rod to another material will cause a current of electrons to flow from the object into the glass rod (since it has a deficit of electrons). Once the rod has been discharged, though, current will cease to flow. We measure current as a flow rate, in units of charge per unit time. An *ampere* (commonly called an *amp,* and abbreviated A) is a flow rate of 1 C/s. If a current of 100 A is flowing (and this is a large current by the way!), that means that 100 C are passing a given point in a conductor every second.

The electric currents that surround us in our daily lives are maintained through artificial means and are able to provide a more steady flow of electrons, a more steady current. We say that any device that maintains a potential difference provides an *electromotive force (emf).* Electromotive force is measured in volts, just like potential difference, introduced in the previous chapter. The batteries that power all the personal electronics that we use, and the electrical generators that provide the current available at the outlets in our homes, provide the potential difference that allows charge to flow.

There are two basic types of current, direct current (DC) and alternating current (AC). *Direct current* refers to electron flow in a single direction with time. Batteries of all varieties provide direct currents. Batteries have a positive and a negative terminal, and are rated in the voltage (potential difference) that they can sustain. Car batteries, for example, are typically 12 V DC. Batteries that power your CD player may be 1.5 V DC, and you may need two of them to provide the power that your CD player requires.

The electrons that flow in *alternating currents* do not push electrons in a single direction but, rather, move back and forth, in a motion similar to the motion of a swing (simple harmonic motion). Because the electrons are constantly changing direction, this implies that the voltage of the emf changes as well. The rate of change of the direction of the current (and voltage) in an AC circuit is measured in cycles per second (1/s) or hertz (Hz). In the United States, the current varies at a rate of 60 cycles per second, or 60 Hz, and maintains a voltage of 110 to 120 V.

RESISTANCE AND OHM'S LAW

Almost all materials resist the flow of current to some degree. Electrons move more easily through some materials than they do through others, as we have seen. Conductors allow electrons to pass more easily than do insulators.

The tendency for certain materials to slow the passage of electrons is called its *resistance.* Resistance is measured in ohms [abbreviated with the Greek letter omega (Ω)] after the German physicist Georg Simon Ohm (1787–1854). The resistance of a material depends on its atomic structure (metals, for example, allow electrons to pass easily through their lattice structure) as well as its temperature. Raising the temperature of materials produces more random motion in their electrons, and this random motion keeps current from flowing easily. Conversely, lowering the temperature of materials can reduce their resistance, allowing current to flow more easily. There are even materials, called *superconductors,* that at sufficiently low temperatures have almost no resistance to the flow of electricity.

An *electrical circuit* is any pathway that allows electrons to flow. Simple circuits can involve very few elements. A flashlight is a simple circuit, involving only an emf (the battery) and a resistor (the bulb). Figure 13.1 shows a diagram of this simple electrical circuit.

Georg Ohm is also credited with discovering the relationship among current, voltage, and resistance in a circuit. He found by experiment that the current in a circuit is directly

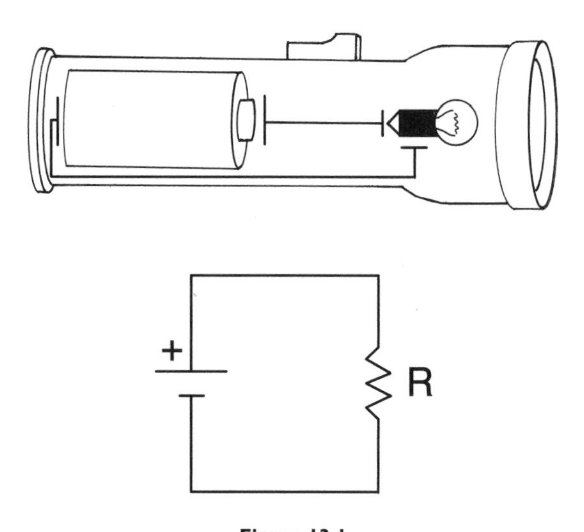

Figure 13.1
A Flashlight and Its Equivalent Circuit

proportional to the applied voltage (the emf), and is inversely proportional to the resistance. We can write this relationship, called *Ohm's law*, as

$I = V/R$

where I is the current (in amperes), V is the voltage (in volts), and R is the resistance (in ohms). Therefore, the unit for resistance, the ohm, is the equivalent of volts per ampere, or $1\ \Omega = 1\ V/A$.

Thus, doubling the voltage of a circuit (while keeping resistance the same) doubles the current. Halving the resistance of a circuit (while keeping voltage the same), also doubles the current.

EXPLORATION 13.1

A circuit consists of a 1.5-V battery and a 1000-Ω (1-kΩ) resistance lightbulb. What current is flowing in the filament of the bulb?

We use Ohm's law to write

$I = V/R$

$I = (1.5\ V)/(1000\ \Omega) = 0.015\ A$, or 15 mA

PHYSICS IN THE REAL WORLD

The more current that passes through a circuit, the more internal heat it can generate. A toaster, for example, contains low-resistance wires that allow a large current to flow through, thus generating enough heat to produce crispy, toasted bread.

ELECTRIC POWER

Power is the rate at which work is done, and is measured in watts, or joules per second. We encountered power earlier in our discussion of work and energy. *Electric power* is the product of current and voltage, or

Electric power = Current × Voltage

or

$P = IV$

For example, let's say that you plug a fixture with a 4-W lightbulb into a 120-V (AC) line. We can determine the current from

$P = IV$

$4\ W = I \times 120\ V$

Therefore,

$I = 4\ W/120\ V = 0.03\ A$, or 30 mA

EXPLORATION 13.2

You can check that the product of current and voltage is, indeed, power by checking units.

Power = Current (charge/time) × Voltage (energy/charge) = energy/time

This type of check (sometimes called unit analysis) is a useful technique in solving problems. If the units of your final answer are not correct, it is quite likely that your final answer is not correct.

Series Circuits

Elements in a circuit can be connected in two different ways, or in a combination of ways. When elements are added in such a way that charge must flow through one element before flowing through another, we say that they are connected in *series*. For example, if two bulbs are connected to a battery as illustrated in Figure 13.2, they are connected in series. When resistors (such as lightbulbs) are placed in a circuit in this way, the total resistance is equal to the sum of the individual resistances, and the current flowing through each resistor in series is the same.

In Figure 13.2, for example, if each bulb has a resistance of 500 Ω, then the total resistance of the circuit is 500 Ω + 500 Ω = 1000 Ω, or 1 kΩ. If the bulbs are connected to a 2-V battery as shown, then we can also calculate the total current to be $I = V/R$, or $I = 2\ \text{V}/1000\ \Omega = 0.002\ \text{A}$, or 2 mA.

The voltage decrease (sometimes called a voltage "drop") across any resistor in series can be determined using Ohm's law, and in general, the voltage drop is proportional to the resistance.

Thus, in the example circuit in Figure 13.2, there is a current of 2 mA flowing through each resistor. The voltage drop across each resistor, then, is determined from

$$V = IR$$

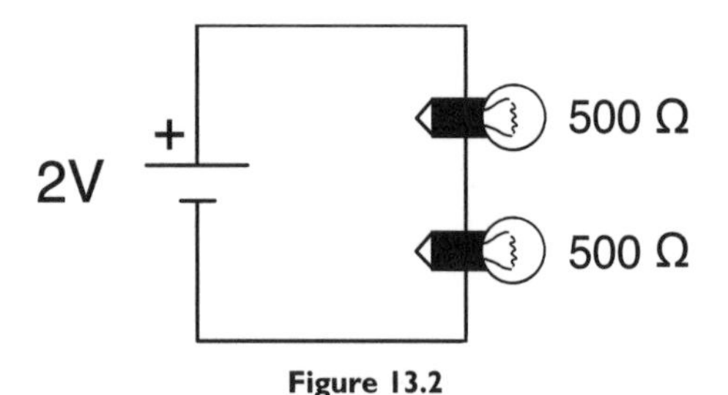

Figure 13.2

or

$$V = 0.002\ \text{A} \times 500\ \Omega = 1\ \text{V}$$

Because in this example all the resistors have the same value of resistance, each has the same voltage drop across it.

Now, imagine a slightly different scenario: If one of the resistors had a resistance of 100 Ω, and the other had a resistance of 900 Ω, the total resistance (in series) would still be 900 Ω + 100 Ω = 1000 Ω, or 1 kΩ. As a result, the current in the circuit would be the same (2 mA), and 2 mA would be flowing through each resistor. However, the voltage drop across the first resistor would be

$$V = IR$$

or

$$V = 0.002\ \text{A} \times 900\ \Omega = 1.8\ \text{V}$$

The voltage drop across the second resistor would be

$$V = 0.002\ \text{A} \times 100\ \Omega, \quad \text{or} \quad 0.2\ \text{V}$$

The total drop in voltage through the circuit is (and must be) the potential difference artificially maintained by the battery, or 1.8 V + 0.2 V = 2 V, but proportionally more of the voltage is "dropped" across the larger resistor.

Parallel Circuits

When resistors are added to a circuit in such a way that current can flow through one or the other resistor, we say that the resistors are in *parallel*. Any devices that are connected to the same two points in a circuit are connected in parallel. The resistances of resistors in parallel add together in a different way from that of

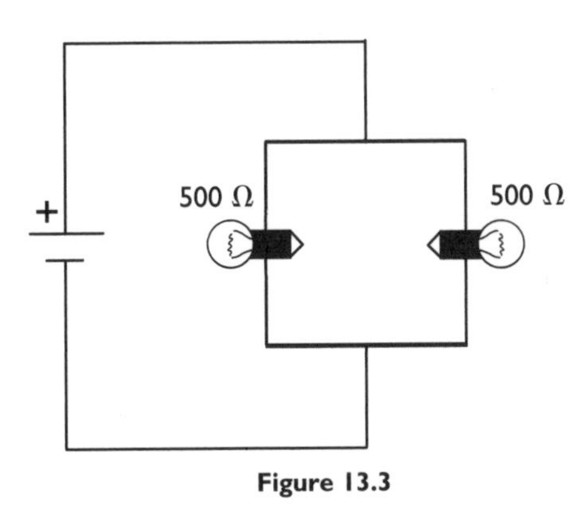

Figure 13.3

resistors in series. The resistances of two resistors in parallel add such that

$1/R_{tot} = 1/R_1 + 1/R_2$

Thus, if the same two bulbs in Figure 13.2 were placed in parallel, the circuit would look as drawn in Figure 13.3. If the resistance of each resistor is 500 Ω, then the total resistance is

$1/R_{tot} = 1/500\ \Omega + 1/500\ \Omega$

or

$1/R_{tot} = 2/500\ \Omega$

so that

$R_{tot} = 250\ \Omega$

When resistors are in parallel, the voltage drop across them must be the same (because they are connected to the same two points in the circuit), but the current flowing through each resistor is now inversely proportional to the resistance of that resistor, according to Ohm's law. Imagine the higher-resistance element in the circuit to be like a restrictive pipe that will not let water pass easily. As a result, the water (current) takes the path of least resistance. Electrical current also takes the path of least resistance.

In the situation where two resistors in parallel have the same resistance, the current through each one will be the same; however, if one resistor has a resistance of 900 Ω and the other has a resistance of 100 Ω, then a larger current will preferentially flow through the "channel" with the lower (100-V) resistance.

EXPLORATION 13.3

If the bulbs are added in parallel, how does the current in the circuit shown in Figure 13.2 (calculated previously) compare with the current in the circuit shown in Figure 13.3?

The series circuit has a current of 2 mA (see previous calculation). The parallel circuit has a current determined by Ohm's law, or

$I = V/R$
$I = 2\ V/250\ V/A$
$I = 0.008\ A$, or 8 mA

Thus, the current in the parallel circuit is higher by a factor of 4, owing to the lower resistance of the same two bulbs placed in parallel.

Resistors can be placed in a circuit in a huge variety of configurations. Most circuits are not simply parallel or series but complicated combinations of series and parallel components.

Figure 13.4a shows a circuit that is a combination of series and parallel resistors. We can calculate the equivalent resistance of this circuit if we simplify it element by element.

EXPLORATION 13.4

We have stated that the potential difference across resistors in parallel is equal, so in our example, each resistor has a potential difference of 2 V across it. What is the current through each resistor if one has a resistance of 100 Ω and the other has a resistance of 900 Ω?

Again, we use Ohm's law to determine the current:

$I = V/R$

$I = 2\ \text{V}/100\ \Omega = 0.02\ \text{A}$ (through the 100-Ω resistor)

$I = 2\ \text{V}/900\ \Omega = 0.002\ \text{A}$ (through the 900-Ω resistor)

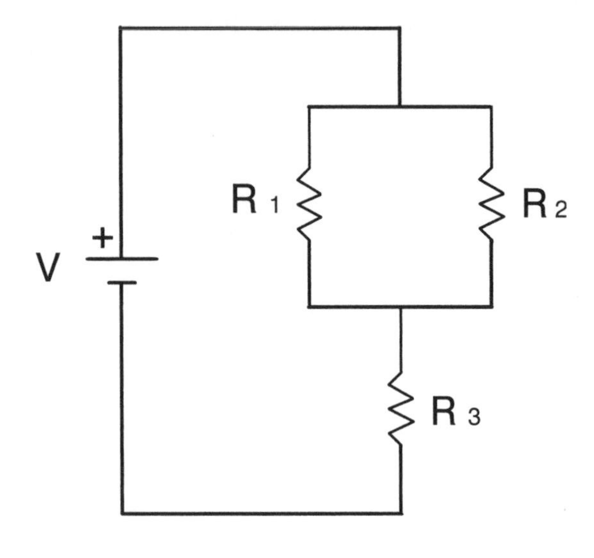

Figure 13.4a

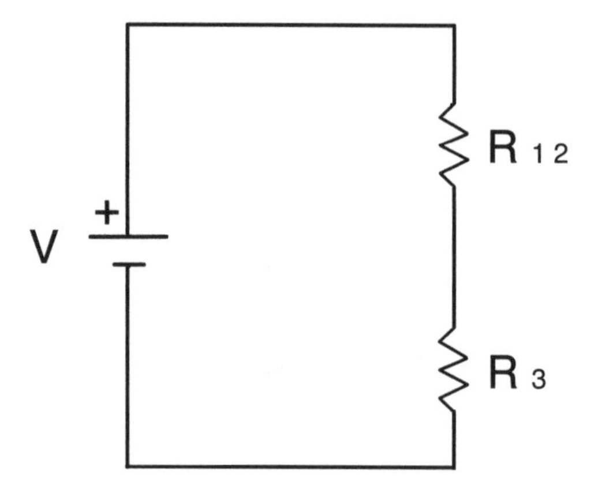

Figure 13.4b

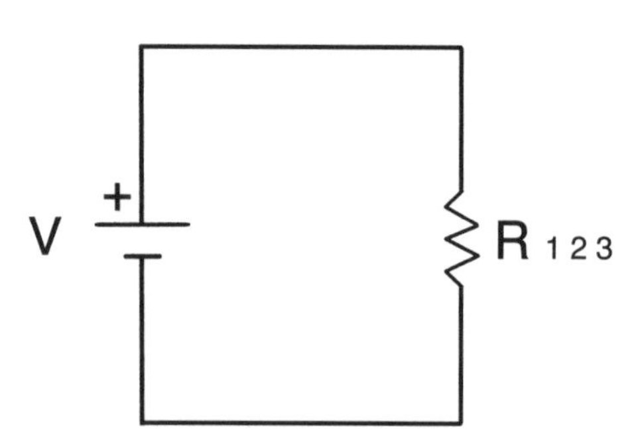

Figure 13.4c

Combining resistors $R1$ and $R2$ (in parallel) gives us a resistance we will call R_{12} (see Figure 13.4b). Equivalent resistor R_{12} and resistor R_3 are in series, so we can simply add their resistances together to get an equivalent resistance that we will call R_{123} (see Figure 13.4c).

Because R_1 and R_2 both have a resistance of 1 Ω, their combined resistance is $R_{12} = \frac{1}{2}\ \Omega$. $R3$ has a resistance of 1.5 Ω. The equivalent resistance $R_{123} = R_{12} + R_3$ (since they are in series). Thus, $R_{123} = 2\ \Omega$.

PHYSICS IN THE REAL WORLD

When bulbs are connected in series and one burns out, then all the lamps will go out, since the path for electrons is blocked by the one broken filament; however, when lamps are connected in parallel, if one lamp burns out, the others will stay lit (and get brighter).

You may have wondered how all the other lights in your string of holiday lights keep burning after one goes out. That fact tells you that the lamps must be wired in parallel.

FUSES AND CIRCUIT BREAKERS

When you plug in electrical devices in your home you are adding them to a parallel circuit. Each device that you plug in draws more current through the wiring in your house; however, if you draw too much current through the wires of your house, you could start an electrical fire. To prevent such a disaster, homes are equipped with fuses that contain a ribbon of material with a low melting point. If the current in the wires exceeds some certain value in amperes, the fuse will "blow," which means that the ribbon of wire will break owing to the heat generated by the current, and the electrical circuit will be broken temporarily.

Alternatively, circuit breakers are commonly used in place of fuses. These devices automatically throw a switch when the current exceeds a preset level. Circuit breakers need only to be reset to restore the flow of current rather than having to replace a fuse. Of course, it is important to determine why fuses are blowing or circuit breakers are tripping, as these could be warning signs of problems in your home's wiring.

CAPACITANCE

The ability of a material to take on more charge as its potential is raised is measured by its *capacitance*. In an electric circuit, a capacitor is an element in the circuit that can store excess charge. A simple capacitor can be built of two flat metal plates separated by an insulator. By preventing charge from moving across the gap between the plates, the insulator raises the capacitance of the capacitor. Capacitance is measured as the ratio of charge to potential difference, or

Capacitance = Charge/Potential difference

The unit of capacitance is the farad (F), where 1 F = 1 C/V.

PROBLEMS

13.1 What is the difference between alternating current (AC) and direct current (DC)? What voltage is found at typical household outlets?

13.2 Why do batteries wear out, or need to be recharged?

13.3 If DC voltages make current flow in a wire, does that mean that there is a net negative charge in all wires? Explain your answer.

13.4 (a) Use Ohm's law to determine the current in a wristwatch that has a 1.5-V battery and a resistance of 150 Ω.
(b) If a 9-V battery is connected to a circuit and the current is measured to be 100 mA, what is the total resistance of the circuit?

13.5 Use Ohm's law to determine the current flowing through your body if you are accidentally electrocuted. You can assume a voltage of 120 V and a resistance of 100 Ω. (Currents of 0.1 A are sufficient to be fatal, because they disrupt the rhythm of the heart.)

13.6 Which of the types of materials discussed in Chapter 12 probably make the best resistors? Would the class of materials known as metals be good resistors? Why or why not?

13.7 What is the total resistance of the circuit shown in the diagram? If the measured current is 10 mA, what is the voltage of the battery?

13.8 What is the total resistance of the circuit shown in the diagram? If the voltage is known to be 9 V, what current will flow through the circuit?

13.9 If a bulb in your string of holiday lights burns out, what should happen to the intensity of the remaining bulbs (assuming that intensity is directly related to the current through each filament)?

13.10 Why is it a bad idea to use plug extenders to plug more than two devices into an outlet in your wall at home?

CHAPTER 14 MAGNETISM AND INDUCTION

KEY TERMS

magnetism, magnets, induced magnetism, magnetic field lines, magnetic poles, right-hand rule, solenoid, electromagnet, electric motor, magnetic flux, generator, transformer

Magnetism refers to the ability of some materials to attract iron. This phenomenon was discovered in naturally occurring rocks in ancient times. Greeks in particular knew that a specific kind of rock from the region of Magnesia could attract iron; these rocks were sometimes called *lodestones*. In ancient China, pieces of such magnetic rocks were used to guide ships after it was discovered that when suspended by a thread, these natural magnets would orient themselves in a north–south direction.

But it wasn't until the nineteenth century—when the Danish physicist Hans Christian Oersted (1777–1851) discovered that electric current and magnetic fields influence each other—that scientists began to appreciate the deep and significant connection between electricity and magnetism. This discovery laid the groundwork for the theoretical work of James Clerk Maxwell (1831–1879), the physicist who outlined what we now call *electromagnetic theory*, which joined together the forces known as *electricity* and *magnetism*. Maxwell first proposed that moving charges are the origin of both electric and magnetic fields.

Thus, much of the machinery of modern life, from electric motors to power plants to transformers, grew out of the ancient discovery of magnetic rocks.

MAGNETS AND HOW THEY WORK

Magnets, or materials that exhibit magnetic properties, are familiar to us in the everyday world. We play with magnets as children, using them to pick up paper clips or nails. The magnetism depends on an intrinsic property of such materials and the electrons in them. Because electrons have charge, and since the motion of charged particles generates both electric and magnetic fields, it logically follows that the orbital motions of electrons within an atom generate tiny magnetic fields. In reality, because the "orbits" of electrons in atom are randomly oriented, the orbital motion of the electron produces little or no net magnetic field, since the many effects of each electron cancel out one another; however, each electron can also be pictured as a tiny spinning top, and this spinning charge generates a magnetic field of its own. In most materials, electrons with opposite "spin" pair up in an atom, so that the net magnetic field is zero; however, in iron and in a few other materials, the fields of these spinning electrons do not completely cancel out, and there is a net magnetic field to each atom.

Neighboring iron atoms can align their magnetic fields in what are called *domains*, and if these domains are all aligned, then the piece

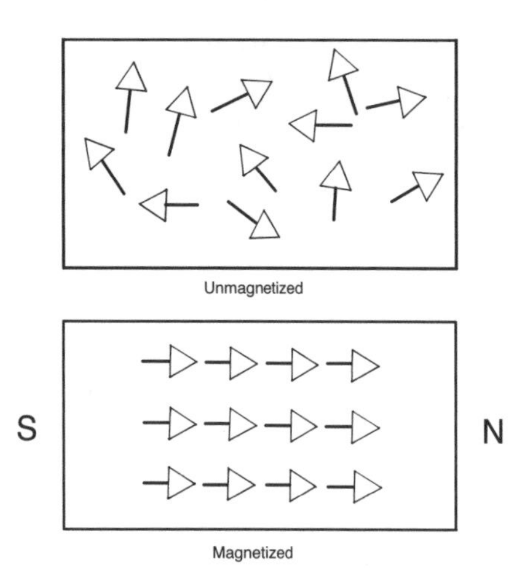

Figure 14.1

of iron is magnetized (see Figure 14.1). If the domains are randomly oriented, then the iron is demagnetized. An ordinary piece of iron or steel can be made magnetic by stroking it—always in the same direction—with either end of a permanent magnet. Besides iron and steel, only a very few other materials can be noticeably magnetized—in particular, the elements nickel and cobalt; but if any magnet is heated red hot and allowed to cool again, it will no longer be magnetic—the jostling of the molecules will have knocked the magnetic domains out of their orderly arrangement.

The theory of these magnetic domains explains what is called *induced magnetism*. We all know that a steel magnet will pick up several tacks or small nails, connected as if they were a chain; however, if the uppermost nail is carefully held in a clamp and the magnet is then removed, the whole chain falls to pieces. Each nail attracted the ones next to it only so long as the magnet was near; we say that orderly magnetic domains were induced in the nails by the permanent magnet. The magnetism of the permanent magnet kept the domains of the iron in the nails lined up temporarily. Iron becomes only temporarily magnetic under the influence of a nearby permanent steel magnet, but a piece of hard steel (like a pin) retains much of its magnetism after the permanent magnet is removed.

EXPLORATION 14.1

If you take a small pile of iron filings and place them on a card and then place a small bar magnet under the card, you will see the iron filings arrange themselves in a certain way. This arrangement results when the iron filings align themselves with the *magnetic field lines* that surround the magnet. Like the electric field lines discussed in Chapter 12, these lines (actually curves) cannot be seen with the eye, but they do exert forces that we can measure, so we know that they are there.

When you rotate the magnet you will notice that the iron filings move with the magnet, indicating that magnetic force is real, capable of doing work like any other force.

The ends of a bar magnet are called its *magnetic poles*. The pole that turns toward the north when the bar is suspended is called a north-seeking or simply the north pole (N); the end that swings toward the south is called the south-seeking or simply the south (S) pole.

The force of attraction or repulsion between poles is directly proportional to the strengths of the magnetic field at the poles and inversely proportional to the square of the distance between them, analogous to the action of the Coulomb force law for electrical force.

If a standard bar magnet is cut or broken in two, each of the new pieces will have a north and a south pole. No matter how many times the cutting process is carried out, the resulting piece of material will always have two poles. There is (thus far), however, no magnetic equivalent to a point charge. No "point poles," or magnetic monopoles, have ever been detected in nature.

MAGNETIC FIELDS

A permanent magnet can exert force from a distance, the forces being exerted through what we call its *magnetic field*. You know this from experience if you played with magnets as a child. If you carefully bring the north pole of one magnet close to the north pole of another, the forces can be sufficient (depending on the mass of the magnets and the friction present) to push the other magnet along. Opposite poles attract, and like poles repel. If one brings one south pole too close to another south pole, one of the magnets is likely to flip around, allowing the two magnets to "stick" together. Children's wooden train sets often employ opposite magnetic poles to hold the train cars together.

Although a magnetic field is strongest near the magnet or magnets that cause it, it extends out indefinitely into space. Within the magnetic field, force is exerted in curved patterns known as *magnetic lines of force*.

We can understand magnetic field lines in relation to the electric fields discussed in Chapter 12. Recall that in electric field diagrams, field lines are drawn from positive charge to negative charge, indicating the direction of motion of a positive test charge. In magnetic field diagrams, field lines are drawn from the north pole (N) to the south pole (S).

Earth's Magnetic Field

Earth itself behaves as if it has an enormous magnet deep within its core. Earth has a North Pole and a South Pole that are the equivalent of the north and south poles of a bar magnet; however, we know that the center of Earth is far too hot to be a magnetized chunk of solid material. Because Earth's core is so hot, its magnetic field is much more likely generated by the motions of ionized material deep beneath its surface. Moving charge generates electric fields, and there is likely some motion in this charged material that generates the magnetic field of Earth.

EXPLORATION 14.2

You can determine the difference (locally) between geographic north and magnetic north if you have a computer and a compass.

At local noon on a sunny day, the shadow of a stick will point toward Earth's geographic north pole. Local noon is defined as the time of day that the Sun transits, or passes, the halfway point in the sky. The Internet site of the U.S. Naval Observatory (http://www.usno.navy.mil) will tell you the time that the Sun will transit at your location. At this time, mark the line of the shadow of a stick perpendicular to the ground. At the same time, and on the same piece of paper, mark the direction that a compass indicates to be magnetic north.

Measuring the angle between these two lines with a protractor will give you the difference between magnetic north and geographic north at your location.

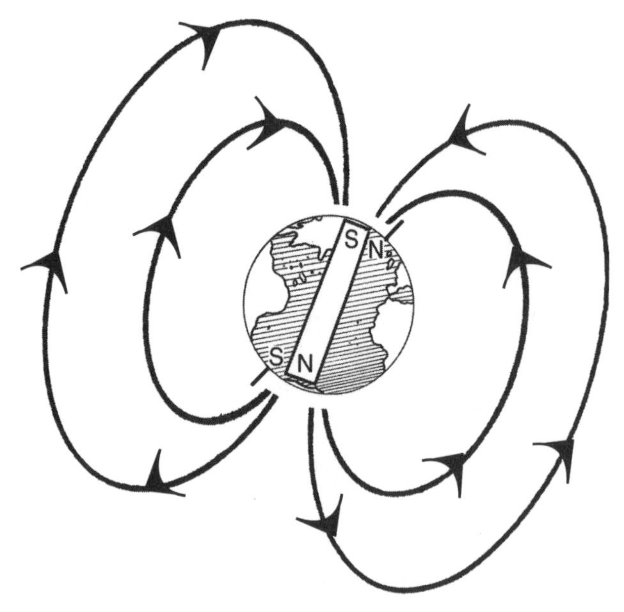

Figure 14.2

Earth's rotational axis passes through what are called its *geographic* north and south poles. The geographic north pole of Earth, for example, points very close to a star in the sky called Polaris, or the North Star. Earth's north magnetic pole is located near its south geographic pole, and its south magnetic pole is found near its north geographic pole (see Figure 14.2).

PHYSICS IN THE REAL WORLD

If you have ever gone hiking or been on a camping trip, you have probably used a compass to figure out the direction that you were heading. Why does a compass needle line up in a north–south direction?

As we have seen, bar magnets have a north and a south pole, and the needle in a compass is a magnetized piece of steel that aligns itself with the magnetic field lines that surround Earth, which run from its magnetic north pole to its magnetic south pole. Thus, the north pole of a magnet swings to the north as it seeks Earth's south magnetic pole.

Because Earth's magnetic and geographic poles are not at the same place, compass-indicated north differs considerably in most places from what is called "true north." In the northeastern part of the United States, a magnetic compass will point north-northwest, whereas on the Pacific coast it will point north-northeast.

ELECTRICITY AND MAGNETISM

Some of the most important technical applications of electricity depend on the fact that a moving charge (current) produces a magnetic field, as discovered by Hans Oersted. He noticed that a compass needle placed just below a wire carrying a current would take up a position nearly perpendicular to the wire when current was flowing. When the direction of the current was reversed, the needle again set itself at right angles to the wire, but with its poles reversed. The effect was detected only while current was flowing and thus could not be due to any magnetism in the copper wire, which is nonmagnetic. In fact, wire is not even required—electrons moving through a vacuum between two charged electrodes are found to have the same effect on a magnet.

The lines of magnetic force that indicate the field due to a current in a straight piece of wire are found to be circles that go around the wire in one direction (see Figure 14.3). The field is strongest near the wire and gets weaker with distance from the wire in any direction. If the flow of current is reversed,

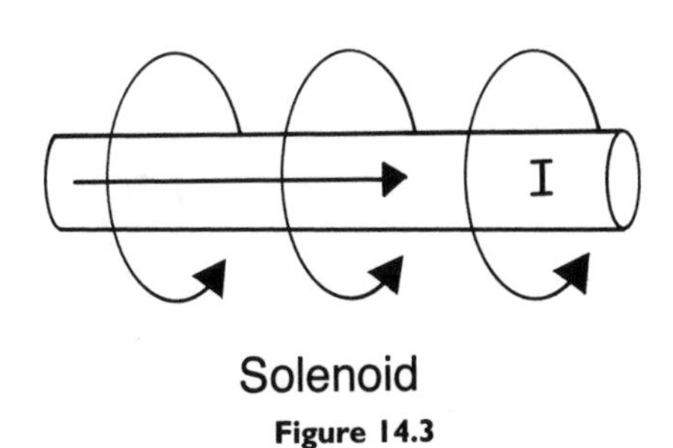

Figure 14.3

the magnetic field lines again circulate around the wire but in the opposite direction. The strength of the magnetic field at some distance r from a wire is found to be directly proportional to the current and inversely proportional to the distance, or

$B \propto I/2\pi r$

From this equation, you can determine that the strength of a magnetic field is measured in units of current over distance, or amperes per meter. You can determine the connection between current flow and the direction of magnetic field lines with the *right-hand rule,* which says that if the thumb of your right hand points in the direction of current flow in a wire, the fingers will curl in the direction of the magnetic field lines that surround it.

ELECTROMAGNETS

The French physicist Andre-Marie Ampére (1775–1836) found that the magnetic field generated by a current in a wire could be greatly increased by coiling the wire into a cylindrical shape. In effect, much more current flows through a smaller region (through each coil in the wire), increasing the magnetic field. Such a coil is called a *solenoid.* The magnetic field surrounding

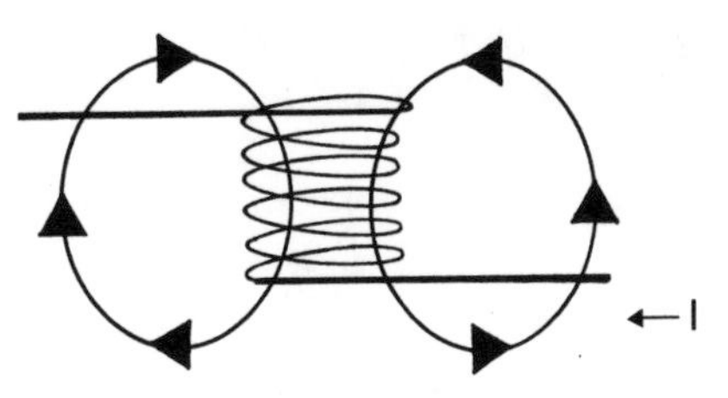

Figure 14.4

the many coils resembles that of a bar magnet. Again, the right-hand rule can be used to determine the direction of the field lines that run through a solenoid (see Figure 14.4).

The strength of the magnetic field generated by a solenoid can be increased hundreds and even thousands of times by placing an iron core inside the coils of the solenoid. This device is sometimes called an *electromagnet* (see Figure 14.5). An electromagnet has many advantages over a permanent magnet, including that its magnetic field is much stronger and that it can be "shut off" by stopping the current.

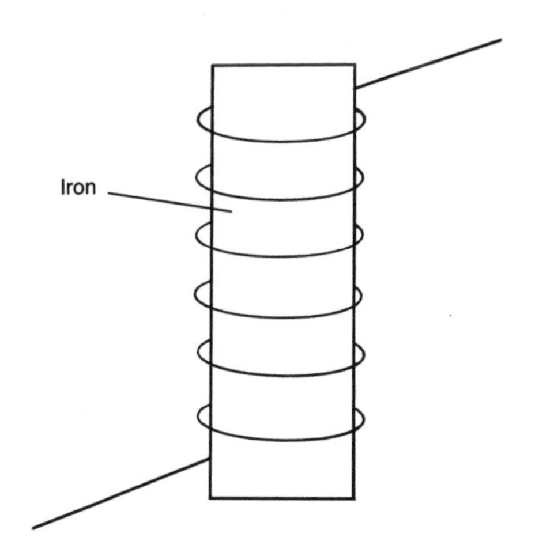

Figure 14.5

PHYSICS IN THE REAL WORLD

If you have passed a junk yard and seen a crane picking up old auto bodies or piles of scrap, you have seen a powerful electromagnet in action. Electromagnets strong enough to lift a weight of many tons are used to load steel rails or bars, machine parts, and scrap iron. The load is picked up or released by closing or opening a switch that controls the current in the coils. Some electromagnets can lift up to 200 lb for each square inch of magnetic surface area.

The speakers in your home or car stereo system are also electromagnets. Current moving through wires exerts changing magnetic forces on magnets attached to a cone of fabric in the speaker. Changes in the current force the speaker to move back and forth, generating sounds waves (compressions) in the air.

Forces Operating on an Electric Current

Experiments indicate that a current-carrying wire placed in a magnetic field experiences a force. As an example, suppose that, in Figure 14.6a, a wire extends in a direction perpendicular to the field lines of a magnet. The direction of the force can be determined using another right-hand rule. If the fingers of the right hand (extended) point in the direction

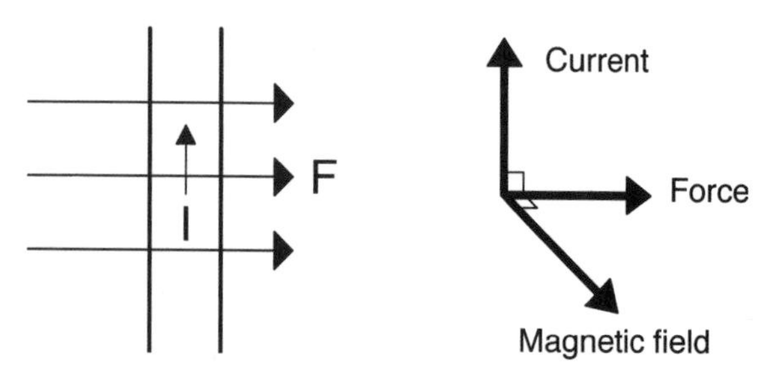

Figure 14.6a

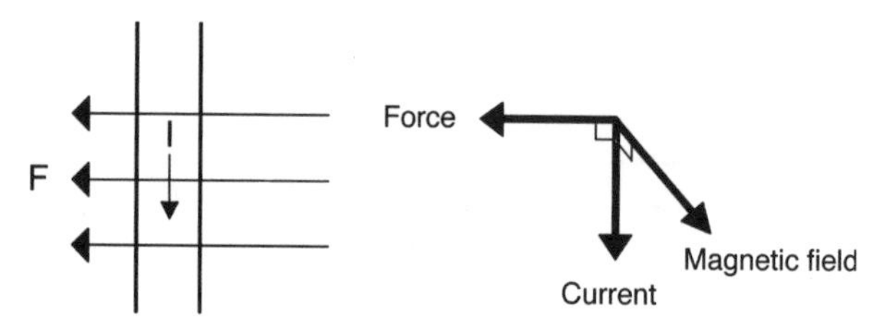

Figure 14.6b

of current flow, and you curl them in the direction of the magnetic field, then the thumb will indicate the direction of the force that the field is exerting on the wire.

With a field and current as shown in Figure 14.6a, the wire is found to be pushed to the right. The three directions—current, field, and force—are perpendicular to one another (see figure inset).

In Figure 14.6b, the direction of the current has been reversed. Using the right-hand rule, you can determine that the direction of the force exerted on the wire is also reversed, and the wire is being pushed to the left.

The strength of the force exerted on the wire is measured to be equal to the product of the charge (q), the velocity of the charge (v), and the strength of the magnetic field (B). Thus, the magnitude of the forces in the example can be calculated to be

$$F = \text{Charge} \times \text{Velocity} \times \text{Magnetic field strength}$$

or

$$F = qvB$$

The formula is useful when written in this form because physicists are often interested in the forces exerted on charged particles that are moving at high velocities and encounter magnetic fields, whether these particles are contained in a wire or not. The Sun, for example, is a source of very energetic charged particles. These particles encounter the magnetic field that surrounds Earth and are deflected with a force that can be calculated using this equation. Fortunately, Earth's magnetic field keeps most of these energetic charged particles from striking its surface.

MEASURING CURRENT AND VOLTAGE

The instruments used to measure current and voltage are *ammeters* and *voltmeters*, respectively. Both ammeters and voltmeters are built around a device called a *galvanometer*, which is a coil of wire mounted in such a way that it is free to rotate in a magnetic field. When current flows in the coil it turns in one direction on its axis. This turning is opposed by a pair of small springs, and since the magnetic force is proportional to the current, the amount that the coil turns is a measure of this current. An ammeter consists of a galvanometer in *parallel* with a resistor (called a *shunt resistor*) with a small resistance, so that it draws most of the current. In the case of a voltmeter, a resistor of high resistance is connected in *series* with the galvanometer. The voltmeter (with a large internal resistance) will not draw much current from the circuit. The movement of the coil in the galvanometer is determined, as described, by the current flowing in it. By Ohm's law, this current is proportional to the applied potential difference, and the scale can be calibrated to be read in volts.

ELECTRIC MOTORS

If a current-carrying coil is allowed to turn freely in a magnetic field, it will acquire kinetic energy. If this turning could be made to continue, it would provide a steady conversion of electrical into mechanical energy. Any device that converts electrical energy to mechanical energy is called an *electric motor* (see Figure 14.7).

For simplicity, suppose the coil consists of only a single loop of wire in the magnetic field of a permanent magnet. Because the current is flowing in opposite directions on the two sides of the coil, the sides of the coil

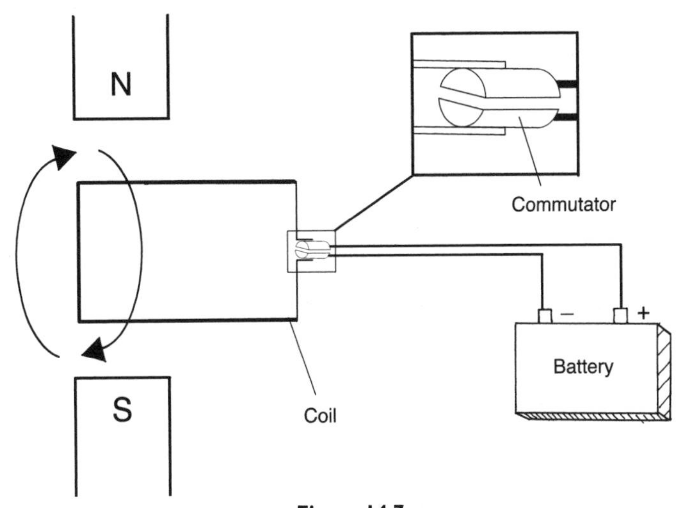

Figure 14.7
A Schematic of an Electric Motor

will be pushed in opposite directions, causing it to rotate. If the coil is free to spin, the current can no longer flow through the coil by fixed wires. To keep the coil spinning, the source of current must be connected to the coil by way of a *commutator,* which is a split ring or cylinder that touches contacts (brushes) attached to the coil. Each half of the commutator is connected to the battery (current source), so the current in the coil will reverse after a half rotation, and the coil will continue to rotate in the same direction.

Let's consider a specific portion of the rotation of the coil (which would be connected to the rotating shaft of a real motor). When the current is flowing through the loop in the direction shown in Figure 14.7, the loop will turn until its plane is vertical. At that moment, however, the current through the loop is automatically reversed by the reversing of connections as the commutator gaps pass the brushes. This reversal lets the coil make another half turn, with the result that the coil turns continuously in one direction.

PHYSICS IN THE REAL WORLD

Commercial motors can convert about 75 percent of their electrical energy into mechanical work. Motors are everywhere around us, from the starter motors that get our car engines started to the motors that drive our electric toothbrushes and refrigerator compressors.

Motors can be driven by alternating current (AC) or direct current (DC). Home computers contain motors that spin the hard drive and drive the cooling fan for the circuit board as well as the central processing unit (CPU) chip.

Electromagnetic Induction

Once it was realized that electricity produces magnetic fields, scientists immediately began looking for ways to produce electric currents by means of magnetism. Almost simultaneously, Michael Faraday in England and Joseph Henry in the United States conducted experiments in which magnets and coils were used to produce electric current. Their discoveries made the commercial development of electricity possible.

PHYSICS IN THE REAL WORLD

New materials have been developed that change shape in response to applied electrical currents. These materials are highly flexible plastics that flex when voltage is applied to them. Known as electroactive polymers (EAPs), these materials are being considered for the development of replacement muscle tissue in paralyzed patients, and also for the moving parts of future robots.

Because EAPs change shape in response to applied current, they are also found to generate electrical currents when they change shape. EAPs, for example, could be placed in the soles of hiking boots to generate current with each step that could charge a battery or power a lamp in a flashlight.

In one experiment, Faraday connected a coil directly to a meter and found that when one pole of a bar magnet was moved quickly toward the coil, the meter registered a momentary current (see Figure 14.8). When the magnet was jerked away, another temporary current was measured, but in the opposite direction. The current seemed to be generated not by the presence or absence of a

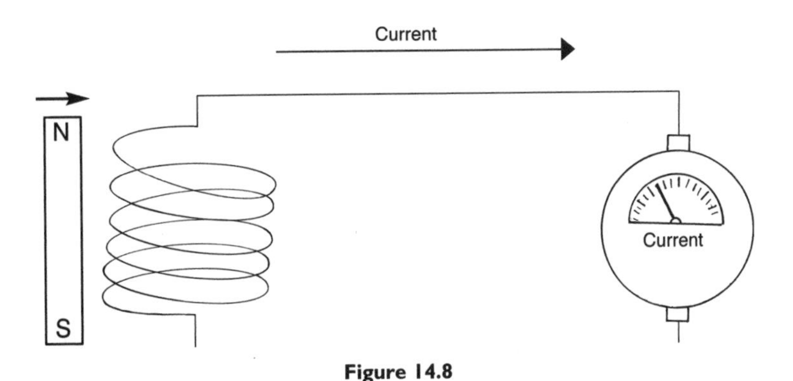

Figure 14.8

field but by the changing strength of the field. Faraday found that the strength of the current increased with the strength of the magnet, its speed of motion, and the number of turns of wire in the coil.

Another experiment showed that an ammeter connected to a coil of wire measured a current at the moment that current started or stopped flowing in an entirely separate circuit nearby. Again, the induced current seemed to result from a changing magnetic field, not its simple presence. As long as a steady current flowed in the primary circuit, nothing further happened, but when the current was stopped in the primary circuit, there was again a momentary impulse of current in the secondary circuit, this time in the direction opposite the original current.

Faraday described what was happening in terms of changing magnetic field. The number of field lines passing through any region of space is referred to as a *magnetic flux*. The SI unit of magnetic flux is called the weber (Wb). One weber is equal to the flux that produces in a single loop of wire an emf equivalent of one volt. Faraday proposed that the current induced in a circuit (which is the equivalent of being momentarily attached to a voltage source or emf) is equal to the rate of change of the magnetic flux. This proposal is called *Faraday's law of magnetic induction.* If the magnetic flux is increasing, then a current is induced in one direction. If the magnetic flux is decreasing, then a current is induced in the opposite direction.

The tesla (T) is the derived unit of magnetic flux density. Magnetic flux density is the measurement of magnetic flux per unit area, and 1 T is equal to 1 Wb per square meter. One tesla is the magnetic field strength necessary to produce a force of 1 N on a charge of 1 C moving perpendicular to the direction of the magnetic field with a velocity of one meter per second.

Experiments have shown that induced currents always produce a field in the direction opposite the magnetic field that induced the current. This must be so, or any induced current would become infinitely large. This generalization is called *Lenz's law* in honor of its discoverer, Heinrich Friedrich Lenz (1804–1865).

Electric Generators

Generators perform the reverse function of an electric motor: they continuously turn mechanical energy into direct or alternating current. They are designed to use electromagnetic induction to produce more than temporary, weak current.

The essential parts of a generator are the same as those of an electric motor: a coil or current-carrying wire, a magnetic field in which the coil can be rotated, and some means for connecting the coil to an outside circuit. In fact, with slight adjustments, the same device may be used as either a motor or a generator. If a current from some outside source is passed into the coil, it rotates and acts like a motor; that is, it converts electrical force into mechanical force. If the coil is mechanically turned, as by an engine or a water-driven turbine, an induced current results; that is, the machine converts mechanical energy into electrical energy.

If the coil of a standard motor is connected to an outside circuit by means of slip rings and brushes, the current furnished to this circuit will be *alternating current* (AC). The number of complete cycles equals the number of coil rotations per second. Thus, an alternating current is the kind that naturally results from the turning of a coil (electromagnet) in a fixed magnetic field. Alternating currents are well suited to many purposes such as heating and lighting.

PHYSICS IN THE REAL WORLD

The coil in a generator can be turned in a variety of ways. Dams collect water and then allow the water to fall from a state of high to low gravitational potential energy. As the water falls it can be made to turn turbines and induce a current. Wind farms are large areas covered in what look like enormous aircraft propeller blades. Winds turn the blades, generating current. Almost any motion, from the back and forth tidal motions at seashores to the pressure generated by heated steam can be converted into electrical energy. One of the challenges facing humans in the future will be to come up with environmentally friendly ways to generate electrical energy.

Other uses of current, such as electroplating or the charging of storage batteries, require *direct current* (DC), which always flows in one direction. AC generators must be modified to generate DC currents.

TRANSFORMERS

We have said that a coil of wire wrapped around an iron core makes an electromagnet, which intensifies the field generated by the motion of the current through the coiled wire. If we place two of these electromagnets next to one another in close proximity (but not touching), we have what is called a *transformer*. One reason that alternating current is widely used to transfer electrical energy over long distances is that voltage and current values may be readily and efficiently changed by the use of these devices.

In principle, the pair of coils in Figure 14.9 is a transformer. Any change in the current in the primary coil induces a current in the secondary coil. If an alternating current is supplied to the primary, there will be a

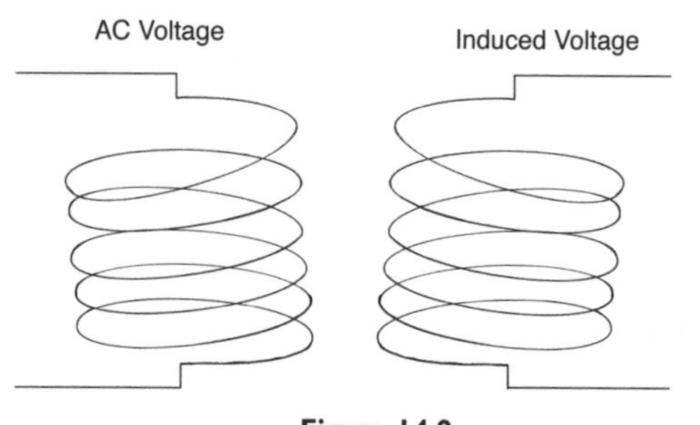

Figure 14.9

corresponding variation of magnetic flux through the secondary. As a result, an alternating current of the same frequency will be induced in the secondary coil. In the United States the frequency used on domestic power lines is 60 cycles per second—that is, the current makes 60 complete oscillations per second, or 60 Hz.

In iron-core transformers, the voltages in the two coils are proportional to the number of turns, or

$$V_s/V_p = n_s/n_p$$

where V_s = voltage in the secondary, V_p = voltage in the primary, n_s = number of turns in secondary, and n_p = number of turns in primary. If there are more turns in the secondary than in the primary, the voltage of the secondary will be greater than the primary voltage, and the device is called a *step-up transformer*; conversely, if there are more turns in the primary than in the secondary, the device is a *step-down transformer*.

When electric power is to be used at a great distance from the generator, it is transmitted in the form of high-voltage AC, for the following reason: the heat loss in an electric power line is proportional to I^2R, so if the losses are to be minimized, the current should be as small as possible. If the power is a constant, this restriction means that the voltage must be high, since $P = IV$.

Transformers work only with alternating currents, as described in the experiments of Faraday and Lenz. In a power plant, the generator voltage may be as high as 10,000 V. A transformer steps this voltage up to perhaps 230,000 V and places this voltage on the transmission line. At the edge of a city, a step-down transformer may reduce the potential difference to about 2300 V, and small step-down transformers located on power-line poles throughout the city then reduce it to a safe value of about 110 V for use in homes.

There are no moving parts in a transformer, and when these devices are properly designed the energy losses may be as low as 2 percent. This means that, practically, the same amount of power is developed in each coil. As in the case of direct current, the power developed in either coil is equal to current multiplied by voltage, so that $I_pV_p = I_sV_s$, or $I_s/I_p = V_p/V_s$. Combining this equation with the preceding relation, we have

$$I_s/I_p = n_p/n_s$$

so that the currents in the two coils are inversely proportional to the number of turns in each.

EXPLORATION 14.3

The primary and secondary coils of a power-line transformer have 50 and 25,000 turns, respectively. Neglecting losses, if AC of effective voltage 110 V is supplied to the primary, what will be the voltage in the secondary?

The relation in the previous formula gives

$$V_s = V_pn_s/n_p = 110 \times 250{,}000/50 = 55{,}000 \text{ V}$$

PROBLEMS

14.1 Will either side of a magnet stick to the metal door of a refrigerator? Explain your answer.

14.2 Draw the magnetic field lines around the two magnets in the diagram (N pole of one facing S pole of another). Is the force attractive or repulsive?

14.3 Draw the magnetic field lines around the two magnets in the diagram (N pole of one facing N pole of another). Is the force attractive or repulsive?

14.4 Most of the power generated in the United States comes from coal-burning power plants in which the chemical energy in coal is converted to high-voltage AC current in a power line. Why is more power not generated in cleaner ways, such as hydroelectrically in power plants associated with dammed rivers?

14.5 An electron is moving to the right with a velocity of half the speed of light in a magnetic flux density of 1 T (down). What is the magnitude and direction of the force experienced by this electron?

14.6 A permanent magnet is slid back and forth inside a coil of wire connected to a circuit (see diagram). Will the bulb light? Why or why not?

14.7 If the secondary current in Exploration 14.3 is 1 A, what will be the primary current?

14.8 Explain why Lenz's law must be true.

CHAPTER 15

THE ATOM AND QUANTUM MECHANICS

KEY TERMS

quanta, quantum mechanics, photoelectric effect, Bohr model of the atom, principal quantum number, de Broglie wavelength, Heisenberg uncertainty principle

The topics and concepts that we have discussed up to this point are areas of physics that have been understood for over a century. Although all the topics we have covered are critical to our understanding of the world, and have relevance to daily life, most of what constitutes modern-day physics is a much more recent undertaking, with less obvious—though just as fundamental—connections to our daily lives.

Starting at the beginning of the twentieth century, a number of discoveries about matter and energy extended our understanding of the universe from its tiniest scales to its most enormous. In the final three chapters of this book we explore the fundamental nature of the atom, the relationship between mass and energy, and a new way to understand time and space in the universe that we inhabit.

THE QUANTUM AND THE NATURE OF LIGHT

Albert Einstein advanced our understanding of the physical world perhaps more than any other human before or since. The fundamental nature of his insights is attested to by the number of times his discoveries and proposals are referred to in these final three chapters. But like any scientist, Einstein was successful only owing to the hard work of others working at the same time, and to the work of the scientists that came before him. A deeper understanding of the fundamental nature of matter and energy took the work of many physicists.

At the turn of the twentieth century Max Planck (1858–1947) first announced that electron energies in the atom appear to be restricted to certain values and that radiation is emitted in discrete packets called *quanta*. The study of the physics of these discrete packets of energy is called *quantum mechanics*.

Experimentally, the energy of electrons was found to be proportional to their frequency ($E \sim f$); specifically, the energy was found to be equal to a constant times their frequency, or

$$E = hf$$

where h is a constant called the *Planck constant*, in honor of Max Planck. Not surprisingly, the Planck constant is a tiny number: $h = 6.67 \times 10^{-34}$ J•s. This constant tells us about the essential "graininess" of the energy carried by light.

A beam of light consisting of many (n) photons carries energy equal to this individual photon energy times the number of photons, or

$$E = nhf$$

EXPLORATION 15.1

Use a magnifying glass to look at an old photograph (preferably black and white) made with a nondigital camera. What you will find is that the photograph is made up of many tiny dots of different shades of gray. The energy deposited by quanta of light (photons) onto the photographic negative changed the chemical structure of molecules on the negative that recorded the image, and that deposition of energy is recorded in the final photograph.

In the late nineteenth century, an experiment with light demonstrated that when some metal surfaces were illuminated, they ejected negatively charged particles (electrons) and produced an electric current. A simplified version of the experimental setup is shown in Figure 15.1. A classical description of this phenomenon, called the *photoelectric effect*, was simply that the incident light caused the electrons to jiggle around, eventually causing them to pop out of the surface of the metal; but several details of the experiment were difficult to explain, including that high-frequency light made the effect easier to observe, and that the rate at which electrons were ejected from the metal seemed to depend on the brightness of the light.

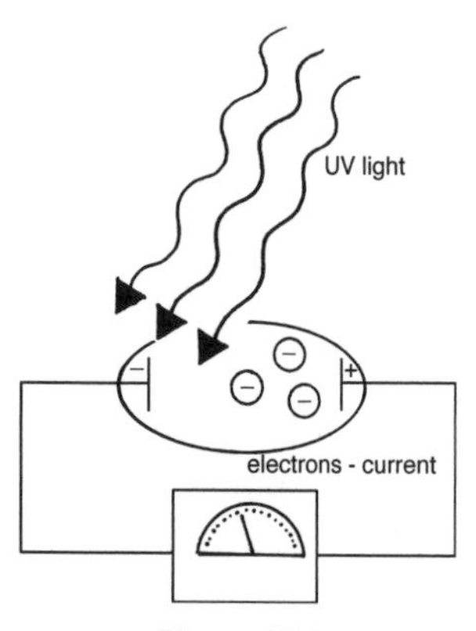

Figure 15.1
The Photoelectric Effect

PHYSICS IN THE REAL WORLD

Digital cameras and many modern electronics depend on the photoelectric effect. When the pixels in the focal plane of a digital camera are illuminated by photons of light, electrons are ejected that generate a current. The rate at which electrons are ejected is proportional to the brightness of the light.

The detector in a digital camera is called a *charge-coupled device* or *CCD*, and these devices are now ubiquitous, appearing not only in digital cameras, but also in cell phones, personal digital assistants (PDAs), and many other modern electronics.

The year 1905 was a good one for the brilliant physicist Albert Einstein, who in that year came up with an explanation of the photoelectric effect and also set forth his special theory of relativity (discussed in the next chapter) that would forever change our perceptions of time. In fact, Einstein won the Nobel Prize for his work on the photoelectric effect, not for his theory of relativity.

Einstein proposed that light can be considered as tiny particles that carry energy (*photons*), and that each time a photon of light with sufficient energy (recall that energy is proportional to frequency) strikes the metal, the metal ejects an electron. If more photons of sufficient frequency strike the metal (brighter light), then more electrons are ejected. If the photons are not energetic enough (their frequency is too low), no electrons will be ejected.

A pure wave model for light cannot explain the photoelectric effect. Half a century after

Maxwell proposed a wave theory of light, Einstein's explanations demonstrated that we still had to consider the *wave–particle duality* of light.

THE STRUCTURE OF THE ATOM

The early years of the twentieth century were also critically important ones for our understanding of the structure of the atom. We stated in previous chapters that the atom consists of a tiny nucleus that contains protons (positive) and neutrons (neutral) surrounded by one or more electrons (negative); however, the exact nature of this structure was not understood at the time that Einstein announced the explanation for the photoelectric effect in 1905. It was known, for example, from experiment that atoms contained positive and negative charges, but the distribution of those charges was not known until the experiments of Ernest Rutherford (1871–1937) soon after.

Using a beam of alpha particles (two neutrons and two protons) from a radioactive source, Rutherford directed the beam through a thin piece of gold foil. A simplified version of the experimental result is shown in Figure 15.2.

What Rutherford found was that although the positively charged alpha particles generally passed right through the gold foil (as expected), occasionally they rebounded as if they had hit a brick wall (owing to the high atomic mass of the gold nucleus). This experiment suggested that the atom is mostly empty space and that the positively charged nucleus of the gold atom is very tiny. The presence of the atomic nucleus was evidenced by the

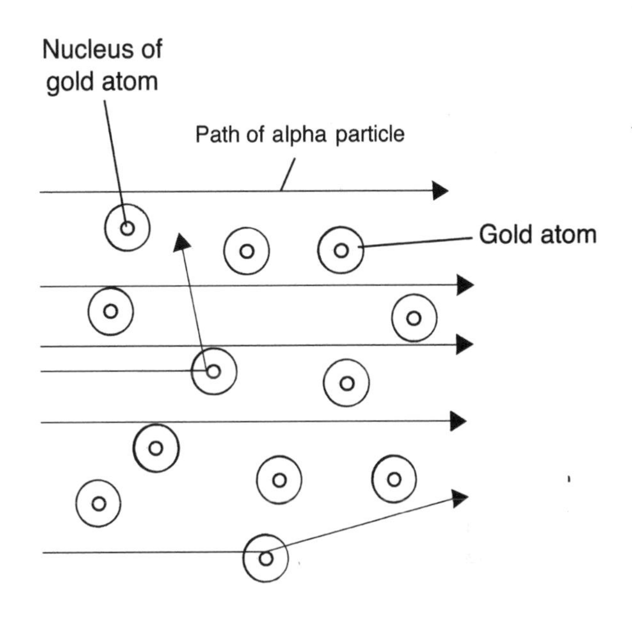

Figure 15.2

rebounding of the positively charged alpha particles when they came close to the positively charged nucleus.

This experiment was the first to confirm that the positive charge was located in a very small volume of the atom and that the electrons were located in some sort of a cloud outside the nucleus.

PHYSICS IN THE REAL WORLD

Rutherford's experiment has been compared to firing a cannon at a piece of tissue paper a large number of times and occasionally finding that the cannon ball bounces back. That is to say, it was a surprising result.

The Bohr Model of the Atom

Neils Bohr (1885–1962) was working at about this same time to understand the nature of spectral lines that chemists had been cataloging during the preceding century. It was well known that elements were found to have unique spectral lines (hydrogen has four distinct lines in the visible part of the spectrum, for example), which served to unambiguously identify the element. Some sample spectral lines are shown in Figure 15.3. Bohr proposed a model of the atom that incorporated the ideas of Planck and Einstein and also explained the origin of spectral lines.

Bohr proposed that the electrons could be pictured as orbiting the nucleus like tiny planets, but that they could exist only in orbits with particular energies. Importantly, Bohr proposed that an electron will not emit radiation in one of these allowed orbits but will emit radiation when it moves between energy levels.

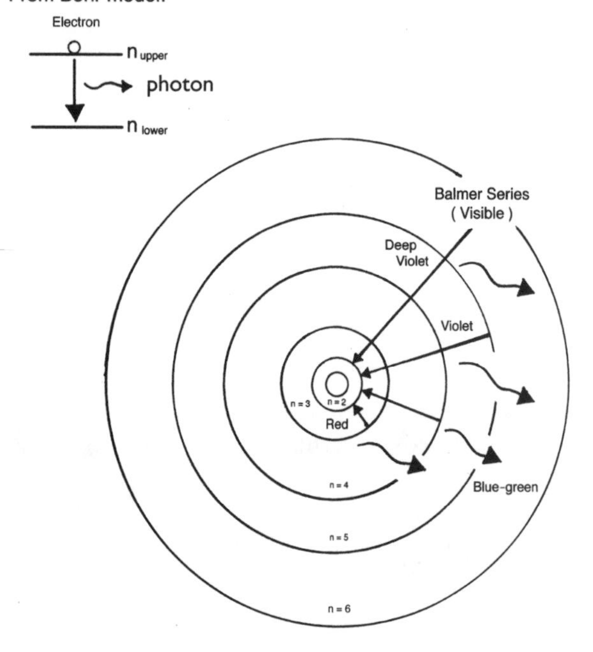

Figure 15.3
The Origin of Spectral Lines from the Hydrogen Atom

EXPLORATION 15.2

Why was it important that electrons not emit radiation when they were in an orbit? If electrons in an orbit emitted energy, then they would gradually lose energy as they orbited. If they lost energy (like a satellite in Earth orbit), they would move into smaller and smaller orbits, pulled toward the nucleus by the attractive nature of the Coulomb force. Eventually, if electrons could radiate energy in an orbit, they would fall into the nucleus of the atom, meaning that even the simplest of atoms (hydrogen, consisting of a proton and an electron) would be unstable. Because we see that hydrogen atoms are stable for long periods of time, Bohr postulated that electrons in allowed orbits could not radiate energy.

The electrons were restricted, as it were, to rungs on a ladder, the rungs being certain distances from the positively charged nucleus. The farther from the nucleus it is, the more energy the electron has. The Coulomb force between the electron and proton keeps the two together, and the motion (angular momentum) of the electron keeps them apart.

Bohr proposed that what was *quantized* in this picture was the angular momentum of the electron (see Chapter 6); That is, the angular momentum of the electron had to have only certain values, and these values were whole-number multiples of Planck's constant (h) divided by 2π. If this were the case, then the electron could absorb or emit photons (packets of energy) as it moved between different energy states, and the amount of energy absorbed or emitted was equal to the difference in the two energy levels.

This model provided another direct link between the energy contained in light (photons) and the energy of electrons. Photons that had the exact right amount of energy—that is, the energy equal to the difference between two energy levels—could be absorbed. All other photons would be unaffected by the atom.

Quantized Energy Levels

In the Bohr model of the atom, electrons in allowed energy levels remain in those levels for a brief time until they naturally fall to a lower energy state.

When they do fall, or "cascade," downward, they give up energy. We have said that energy of a photon corresponds to a particular frequency, $E = hf$.

Because frequency, wavelength, and the speed of light are related by

$$c = f\lambda$$

we can write the expression for the energy of a photon as

$$E = hc/\lambda$$

indicating that the energy of a photon is inversely proportional to its wavelength. Long-wavelength photons (e.g., radio photons) carry less energy than short-wavelength photons (e.g., gamma rays).

The lowest-energy state, or *ground state,* is the lowest possible energy state that an electron bound to an atom can have, and is assigned the *principal quantum number* $n = 1$. Higher energy levels have principal quantum numbers $n = 2$, $n = 3$, and so on.

There is no highest principal quantum number for an electron bound to an atom; however, if enough energy is absorbed by an electron, it will be entirely removed from the nucleus. We call this removal of the electron from the atom *ionization*. The Bohr model of the atom can be used to calculate the ionization energies of different elements.

EXPLORATION 15.3

As an example, in the hydrogen atom, when an electron moves from the $n = 3$ to the $n = 2$ energy levels, it gives off a red-wavelength photon with a wavelength of 656.3 nm. What is the energy separation of these two energy levels?

This wavelength can be used to calculate the energy difference between the $n = 3$ and the $n = 2$ energy levels.

Since

$$E = hc/\lambda$$

we simply use the values for h, c, and λ to determine that

$$E = (6.67 \times 10^{-34}\ \text{J s})(3 \times 10^{8}\ \text{m/s})/(656.3 \times 10^{-9}\ \text{m})$$

or

$$E = 3.05 \times 10^{-19}\ \text{J}$$

As described earlier, an electron-volt (eV) is a very tiny unit of energy used in discussing electron energy levels and is equal to 1.602×10^{-19} J. Thus, our answer to this question can also be expressed as 1.9 eV.

PARTICLES AS WAVES

Neils Bohr and others realized that the model of an electron as a tiny planet orbiting the nucleus could not be physically correct for reasons of stability, although it was a model that allowed physicists to calculate many observable properties of atoms and—more importantly—to explain their spectral lines.

Another physicist, Louis de Broglie (1892–1987), had an explanation of the structure of the atom that was quite radical and more fundamental. He proposed that in the same way that light had particle properties, matter could have wave properties. He proposed in the 1920s that each particle has a wavelength that is related to its momentum (p) by the relation

$$\lambda = h/p$$

where h is again Planck's constant; this wavelength is called the *de Broglie wavelength*.

What de Broglie suggested about electron energy levels was that the electrons that are bound to a nucleus can be considered to be like waves, and that the *allowed* energy states that Bohr proposed are not arbitrary but are in fact the equivalent of standing waves, like the standing waves on a string.

Electrons can be thought of as waves that surround the nucleus in certain orbits that have a circumference; and electrons can be located in orbits where the circumference of the orbit is an even multiple of an electron wavelength.

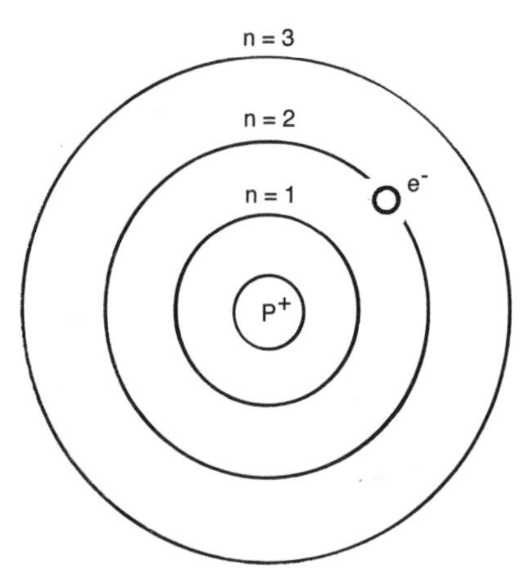

Figure 15.4
Bohr Model of the Atom

Electrons do not collapse onto the nucleus because the ground state corresponds to a standing wave of a single wavelength around the atomic nucleus. Figure 15.4 explains this model of electron orbits.

SCHRÖDINGER AND QUANTUM MECHANICS

Erwin Schrödinger (1887–1961), a theorist also working in the 1920s, gave mathematical support to the ideas of Bohr and de Broglie by proposing a *wave equation* that could describe the electron waves that surround the nucleus. What he proposed was that the location of the electron around the nucleus was not definite but involved *probabilities*.

Using the Schrödinger equation, one can describe the likelihood of finding an electron in a certain orbital but cannot describe exactly where the electron is located. The electron orbits that chemists use to predict the ways in which compounds will form are calculated using Schrödinger's wave equation.

Albert Einstein never liked this probability approach to understanding the atom, and famously stated that, "God does not play dice [with the universe]."

Heisenberg Uncertainty Principle

The concept of uncertainty as it relates to the atom was first explored fully by Werner Heisenberg, who proposed that we can measure precisely *either* the position *or* the momentum of a particle very well, but we cannot know both with infinite accuracy. It is as if there is a smallest marking on the scales of the universe, and that smallest marking is described by the Planck constant.

In Heisenberg's work, the product of the uncertainty in the position and uncertainty in the momentum of a particle is found to equal a constant, the Planck constant, divided by 2π. Mathematically, this relationship, known as the *Heisenberg uncertainty principle*, is stated as

$$\Delta p \, \Delta x \geqslant h/2\pi$$

where Δp is the uncertainty in momentum of the particle and Δx is the uncertainty in its position. Because h is a constant, this equation means that the more certain we are about the momentum of a particle (Δp gets small), the less certain we must be about its position (Δx gets large).

The same sort of relationship seems to be true for the uncertainty in the energy of a particle (ΔE) and the time required to measure that energy (Δt). This uncertainty is expressed as

$$\Delta E \, \Delta t \geq h/2\pi$$

This second uncertainty principle explains one of the properties of the spectral lines that elements emit. Recall from our discussion earlier in the chapter that when an electron moves from a higher energy level to a lower energy level, it gives off a photon of light that has a particular energy, and since $E = hf$, a particular frequency.

These escaping photons (with their particular frequencies) give rise to the spectral lines that we observe to be unique to various elements and molecules.

If the uncertainty principle were not true, then spectral lines would be infinitely "thin," occurring at exactly a single frequency; but because there is an uncertainty in the energy of the photon (related to the uncertainty in the time required for an electron to go from a higher energy level to a lower one), the spectral line associated with that photon has a "natural width" that can be determined.

PROBLEMS

15.1 Why do physicists think that energy in the universe is quantized?

15.2 What is the energy of a single photon with a wavelength of 2 cm?

15.3 What is the energy of a single photon with a wavelength of 10 μm?

15.4 Why can the model of "planetary" electrons orbiting a nucleus not be physically correct?

15.5 What is the lowest principal quantum number of an atom? What it the highest principal quantum number?

15.6 What does it mean to say that an atom is ionized?

15.7 What is the wavelength of the photon generated by an electron moving from the $n = 2$ energy level to the $n = 1$ energy level if its energy is 10.1 eV?

15.8 How is the Schrödinger description of the atom fundamentally different from any other that had come before?

CHAPTER 16

RELATIVITY

KEY TERMS

ether, rest frame, special relativity, time dilation, twin paradox, length contraction, Lorentz factor, general relativity

PHYSICS IN THE REAL WORLD

Imagine two clocks, one on the ground floor of an office building and one on the 40th floor. As Earth rotates, the clock on the top floor is moving farther in the same amount of time. If you imagine the circle that the clock on the top floor traces out in space, it is a larger circle than the clock on the ground floor, yet the clocks travel these different distances in the same amount of time. What does this mean? It means that the clock on the 40th floor is moving at a higher velocity.

According to the theory of special relativity, since the clock on the 40th floor is moving at a higher velocity, it actually measures that time passes more slowly—an effect known as time dilation. Thus, if you want to remain eternally youthful, work on the top floor of your office building.

In our discussion of motion in the early chapters of this book, we recognized that motion must be measured in reference to some fixed object. Thus, a car moving at 60 mi/h has a velocity relative to the surface of Earth. The surface of Earth, however, is rotating once every 24 hours, and the planet itself is orbiting the Sun once every 365.25 days. The motion of the car is not 60 mph in those contexts. We measure velocities relative to some fixed rest frame.

Once it was recognized that light had wave properties (as early as the time of Newton), considerable thought was dedicated to determining the "substance" through which light waves move. Sound waves move through air, water waves move through the ocean, and waves can also move on a string. Thus, typically, one can identify the medium, or substance, through which a wave is moving.

It was hypothesized that light must move through a substance that was called *ether*, and the ether would constitute a frame of reference with respect to which light moved. This would imply that there was a frame of reference in which one could measure the speed of light to be other than 3×10^8 m/s. If you are driving on a freeway at 30 mi/h, and a car going 60 mi/h passes you, it is moving at a relative velocity of 30 mph. Because light has such a high velocity, it was known that this effect would be difficult to measure.

A. A. Michelson (1852–1931) and E. W. Morley (1838–1923) devised an experiment that would allow them to measure the motion of Earth through the ether. Their apparatus measured the speed of light using a device called an *interferometer*, which measures the interference between two beams of light.

The Michelson–Morley experiment found *no* evidence that the speed of light varied as Earth moved through the ether. In the experiment, the orbital speed of Earth was known to be about 30 km/s. Whether the light moved parallel to Earth's motion around the Sun (presumably giving it an extra kick in velocity) or perpendicular to Earth's motion around the Sun, there was no difference in the measured velocity of light, even though the setup was sensitive enough to detect such a change in velocity if it occurred.

EXPLORATION 16.1

What percent change in the velocity of light should the Michelson-Morley experiment have detected?

The speed of light is 3×10^5 km/s, and the speed of Earth in its orbit is about 30 km/s. Therefore, the level of the shift would be

30 km/s / 300,000 km/s = 3×10^{-4}, or 0.0003, or 0.03%

As a result of this famous experiment, physicists concluded that there did not appear to be any *rest frame* for light; that is, there was no frame of reference with respect to which the speed of light could be measured. But no one quite knew why.

SPECIAL RELATIVITY

The experimental results about the ether led Albert Einstein (1879–1955) to arrive at some fundamental postulates about the nature of the universe. These postulates form the basis of what is called *special relativity* and were first proposed by Einstein in 1905.

The observation that there was no universal frame of reference led Einstein to the *first postulate of special relativity*, which is that the laws of nature are the same in all nonaccelerating frames of reference. What does that statement mean?

Here is a practical example. Imagine that you are on a train traveling across the country at night. You are moving at a constant speed of 100 mi/h, and the motion is perfectly smooth, because we will imagine you are in a magnetic levitation (mag-lev) train.

Now, picture that the shades are drawn in your compartment, and you can't see outside. According to the first postulate of special relativity, there is no experiment that you can devise to tell if you are moving at some constant velocity or are at rest. If you pour water into a glass on your train trip, it will flow as if you were at rest. Moving at a constant 100 mi/h is a nonaccelerating frame of reference with respect to the surface of Earth, just as being at rest with respect to the surface of the Earth is. Therefore, the laws of nature behave equivalently in these two situations.

The second postulate is related to the significance of the observations of the Michelson–Morley experiment. The *second postulate of special relativity* can be stated that the speed of light is constant in the universe, regardless of the motion of the source or of the observer.

Although photons of light may have both wave and particle properties, they are not like any waves or particles that we have studied thus far. This postulate is precisely what Michelson and Morley observed at the turn of the twentieth century. There was no rest frame relative to which the speed of light can be measured; all observers measure the speed of light to be the same.

The second postulate might seem relatively simple on its face, but it has some very far-reaching consequences. The fact that the speed of light must be a constant for any observer (even an observer moving relative to another one) means that the two quantities from which we derive velocity (distance and time) must be more pliable than we thought.

PHYSICS IN THE REAL WORLD

If you are accelerating (e.g., speeding up, slowing down, rounding a corner), then Einstein's postulate no longer holds. In those situations, the laws of nature appear to change.

Have you ever tried to play catch with someone on a merry-go-round? Circular motion is a form of acceleration (constantly changing direction), and playing catch on a merry-go-round is like living in a world with strange gravity. In the frame of reference of the rotating merry-go-round, tossed balls appear to move in curved paths. An observer to the side of the merry-go-round, however, will see the ball moving in a normal arc but will see that the receiver is moving (see diagram).

Time Dilation

One of the dramatic effects of the second postulate of special relativity has to do with the perception of time. What follows is sometimes referred to as a "thought experiment," because it illuminates an experiment that would be difficult to carry out physically.

This thought experiment involves a simple apparatus: a light source and a mirror that is

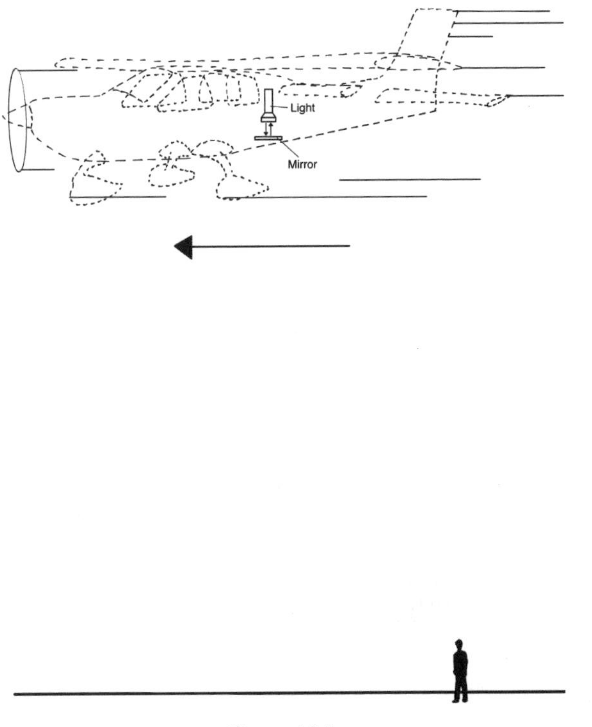

Figure 16.1

contained in a glass airplane flying past a stationary observer (see Figure 16.1). The glass plane is moving at constant velocity relative to an observer on the ground. The observer in the airplane sees that the light travels from the light source, bounces off the mirror, and then travels back to its origin, traversing a distance that is twice the distance between the mirror and the light source.

The observer outside the plane sees something different. She sees the light travel on a diagonal path, strike the mirror, and then travel back on a diagonal path to the light source. She sees the light traverse a longer distance, and the length that she sees the light travel is related to the velocity of the glass airplane. If the glass airplane were at rest with respect to the stationary observer, the two observers would agree on how far the light had traveled.

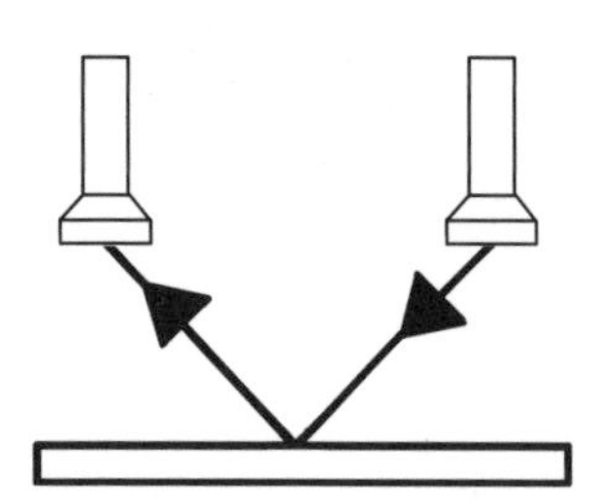

Figure 16.2
Light Path as Seen by the Stationary Observer in Figure 16.1

Figure 16.2 shows a detail of the path of the light as seen from the stationary observer. Both observers (according to the second postulate of special relativity) will agree that the light travels at c, the speed of light; but because velocity is distance traveled over time, the two observers are going to be in a bit of a fix. They agree on the velocity but disagree on the distance. Since

Velocity = Distance/Time

they are going to disagree on how much time has passed. Because the observer at rest saw the light travel a longer distance, she will claim that more time passed. The observer in the plane will say that less time passed.

Specifically, the observer at rest will claim that the time that has passed is

$$t = t_0/\sqrt{[1 - v^2/c^2]}$$

where t_0 is the time observed by the pilot of the plane, v is the velocity of the plane (the relative velocity between the two observers), and c is the speed of light.

Notice that if the speed (v) is much less than the speed of light (sometimes written $v \ll c$), then this particular effect, sometimes called *time dilation*, will be very difficult to detect. For example, the difference in time measured on the ground floor and the 40th floor of an office building is small but measurable with very precise clocks. *Time dilation* refers to the fact that an observer at rest will observe that more time has passed than an observer who is moving at high velocity. Like all effects in special relativity, time dilation becomes significant only when velocities are close to the speed of light.

The Twin Paradox

Time dilation gives rise to a scenario that is sometimes referred to as the *twin paradox*. Special relativity predicts that if one twin leaves on a voyage into space on a particular day, moving at an appreciable fraction of the speed of light to his destination and then returns from the destination, more time will have passed for his homebound twin than for him. He will return to Earth having aged less.

If humans develop the technology to travel safely at these high velocities (still a far in the future possibility), then humans could conceivably travel to the nearby stars within a human lifetime. The problem is that they may return to a very different world, one in which hundreds of years have passed, a world that may have indeed long since forgotten about them. The possibilities for science fiction plots are endless!

Length Contraction

A parallel to time dilation is *length contraction*. In fact, it is possible to think of the twin paradox in terms of length contraction. Objects moving close to the speed of light will appear to an outside observer to contract in the direction of their motion. Thus, the space between the astronaut and his destination contracts as he travels at close to the speed of light toward it. This way of thinking about length contraction emphasizes that special relativity leads us to conclude that high relative velocities between observers lead to a distortion of space and time, not a distortion of objects themselves.

EXPLORATION 16.2

If one traveler leaves Earth and travels for 10 years at half the speed of light, stays at her destination for 10 years, and then travels back for another 10 years (according to her clock) at half the speed of light, 30 years will have passed according to her biological clock. How much time will have passed on the planet that she left?

First, on the trip to the distant star, 10 years pass for her, but on the home planet (at rest relative to her), the time is measured to be

$$t = 10y / \sqrt{1-(0.5)^2/c^2}$$

$$t = 10y / \sqrt{1-0.5^2}$$

$$t = 11.6\ y$$

Then, 10 years pass while the traveler is on the distant planet and, finally, another 11.6 years on the return trip.

Thus, when the traveler gets home, she will be 30 years older, whereas everyone she left behind will be 33.2 years older. The magnitude of the difference gets larger as the velocity of the traveler relative to the person at rest approaches the speed of light.

RELATIVISTIC MOTION AND ENERGY

That light cannot travel faster than c means that nothing in the universe can move faster than c. In effect, the universe has a speed limit.

Photons can move at the speed of light because they have no mass. Objects with mass, much to the dismay of science fiction authors, require a lot of energy to move at very high velocities because the mass of an object moving at velocity, v, is given by

$$m = m_0/\sqrt{(1 - v^2/c^2)}$$

where m is the mass measured by an observer at rest, and m_0 is the *rest mass*, or the mass of the object were it not moving relative to the observer.

This means that as an object with mass travels at a speed close to c, its mass increases; and if an object with mass were to move at the speed of light, it would have (impossibly) an infinite mass.

You may have noticed that the same factor appears in the equations for time dilation, length contraction, and now the relativistic mass. This factor is sometimes abbreviated with the Greek letter gamma (γ) and is called the *Lorentz factor*:

$$\gamma = 1/\sqrt{(1 - v^2/c^2)}$$

Objects moving at 90% of the speed of light have a Lorentz factor of about 2, and objects moving at 95% of the speed of light have a Lorentz factor of about 4.

THE MOST FAMOUS EQUATION

In what is perhaps the most famous equation in physics, Einstein in 1905 proposed that mass and energy are equivalent. As he put it: the mass of a particle is just a measure of its energy. More specifically, Einstein proposed that

Energy = Mass × (Speed of light)^2

or

$E = mc^2$

What this equation means is that each kilogram of material contains an enormous amount of energy (calculated as it is from the square of the speed of light, already a large number).

This equivalence of mass and energy explains how the Sun is able to generate the enormous amounts of energy that it does for billions of years. Through the processes of nuclear fusion (see Chapter 17), a tiny amount of mass can be converted into enormous amounts of energy.

GENERAL RELATIVITY: A NEW THEORY OF GRAVITY

We have said that the special theory of relativity tells us about the physics of reference frames that are at rest or moving at constant velocity with respect to one another. After proposing this theory, Einstein worked to generalize relativity to *accelerating* reference frames.

Fundamental to this general relativity is the idea that gravity and an accelerating reference frame are indistinguishable by any experiment that one might devise. In fact, the *principle of equivalence*, part of the theory, states this idea: experiments cannot distinguish between an accelerating reference frame and a gravitational field.

PHYSICS IN THE REAL WORLD

The principle of equivalence is helpful when one is planning interstellar travel. If a spacecraft accelerates at g (9.8 m/s/s—or the gravitational acceleration at the surface of Earth), astronauts on board the spacecraft will experience a gravitational "field" that is the same as the gravitational field at the surface of Earth.

If the spaceship turns 180° halfway to the distant star and decelerates for the second half of the trip (again at g), then the astronauts can experience an artificial gravitational field for the entire trip.

Newton proposed that gravity was an attractive force between any two objects with mass, and that the force fell off with distance squared. Einstein proposed a very different idea.

Let's engage in one more thought experiment. Imagine that a beam of light is directed across the interior of an accelerating rocket (see Figure 16.3). To an outside observer, the

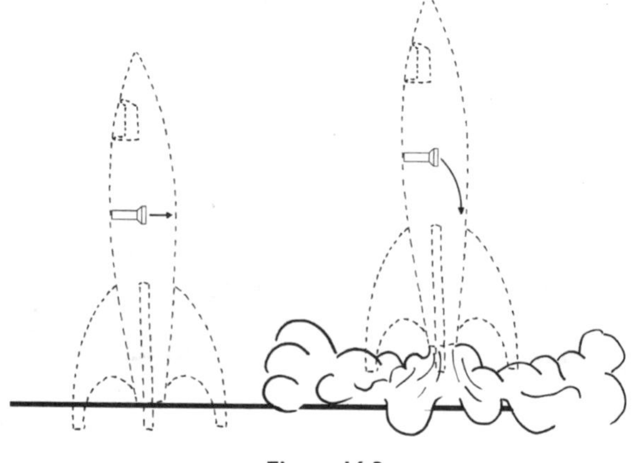

Figure 16.3

beam of light will appear to move in a straight line. But to the observer in the rocket (moving in an accelerating reference frame), the light will appear to strike the opposite side of the rocket lower than the flashlight. In fact, since the rocket is accelerating up, the beam will appear to bend in a parabola before striking the far side.

In effect, the beam of light will appear to be affected by the acceleration of the rocket, and according to the principle of equivalence, one cannot devise an experiment to differentiate between a gravitational field and acceleration.

As a result, this means that we must accept that the paths of photons can be deflected by gravitational fields. Newton had no explanation for this phenomenon (since he postulated that gravity was an attractive force between two masses). Einstein, however, proposed that mass (or its energy equivalent) serves to distort space, and other masses (and photons) move in curved paths because of the distortion.

According to general relativity, matter can act as a gravitational "lens," magnifying distant objects. The wisps of light shown here are the distorted light from a background source located far beyond the foreground cluster of galaxies.

As strange as the general theory of relativity may seem, it has been verified by observation many times since it was proposed. The first confirmation came in 1919 when Einstein correctly predicted the bending of the light rays from a star as they passed close to the mass of the Sun during a solar eclipse.

In the final chapter we turn to an examination of the very large and the very small and the deep connections between them.

PHYSICS IN THE REAL WORLD

Astronomers have discovered objects in space known as *gravitational lenses*. These are concentrations of mass that are so great that they bend the light of the galaxies that are farther away than themselves (see Figure 16.4).

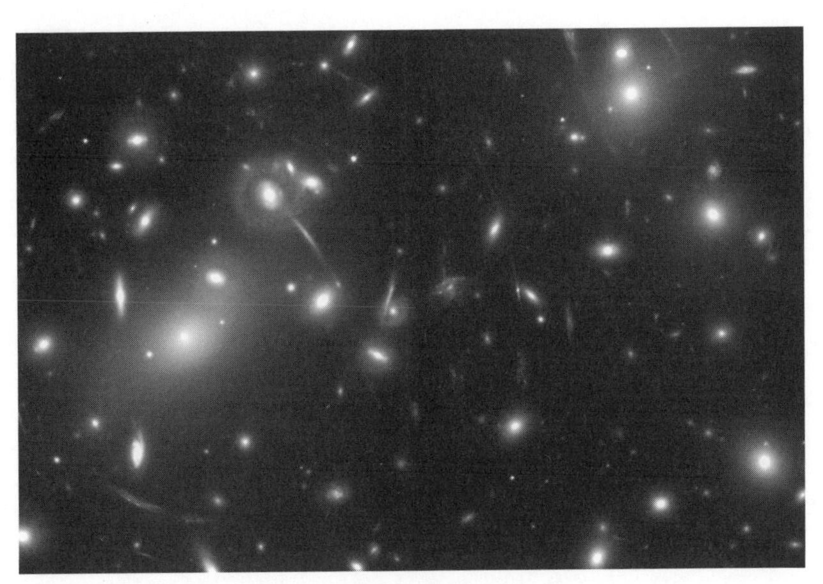

Figure 16.4
A Gravitational Lens [Photo courtesy of NASA, A. Fruchter, and the ERO Team (STScI and ST-ECF).]

PROBLEMS

16.1 Why was the Michelson–Morley experiment critically important to Einstein's proposal of special relativity?

16.2 What is the Lorentz factor experienced by an electron moving at 98% of the speed of light?

16.3 What is the increase in the mass of an electron moving at 95% of the speed of light? (*Hint:* Just solve the mass equation for m/m_0.)

16.4 The star nearest to Earth is Proxima Centauri, located at a distance of about 4.2 light years. If one could travel there at 99% of the speed of light, what would be the apparent distance to the star?

16.5 If you traveled to Proxima Centauri at this high velocity, stayed there for 2 years, and then returned, how much time would have passed on Earth?

16.6 Use Einstein's mass–energy equivalence to determine the energy (in joules) that a typical human (mass = 75 kg) contains.

16.7 Use the mass–energy equivalence equation to determine the energy (in joules) that a proton contains.

16.8 What is "general" about the general theory of relativity?

NUCLEAR PHYSICS, PARTICLES, AND THE COSMOS

KEY TERMS

x-rays, radioactive decay, alpha particles, beta particles, gamma rays, nucleons, atomic number, isotopes, atomic mass number, half-life, strong nuclear force, weak nuclear force, fission, fusion, Big Bang

We end our journey through physics with a chapter that connects the very small to the very large. At the small end, we will consider the constituents of atoms, the building blocks of those indivisible particles first proposed by Greek thinkers so long ago. In particular, we will look at the structure of the nucleus of the atom and how the universe makes elements from its raw materials. At the large end of the spectrum, we will consider the origin of the universe in the Big Bang and how conditions in those early moments of the universe determined all else that followed.

THE ATOMIC NUCLEUS

We discussed during our journey the Coulomb force, the electrostatic force that acts between charged particles, repelling like charges and attracting opposite charges. We said that the negatively charged electron is held to the nucleus by its positive charge. But what holds the nucleus together? One might think that the nuclei of atoms, composed of one or more protons would be unruly objects, trying to fly to bits all the time, with protons repelling other protons.

In fact, the nuclei of atoms do fall apart, sometimes on their own and sometimes with help from outside. The discovery of photosensitive chemicals in the nineteenth century led to the development of photography, with profound effects on the arts and the sciences. Finally, astronomers had a way to archive images of the nebulae, stars, and spectra that they were observing. And at the turn of the twentieth century, German physicist Wilhelm Roentgen (1843–1923) discovered a kind of ray that was invisible to the eye, could pass through some materials, and would expose photographic film. He called these mysterious rays *x-rays*, and we now know *x-rays* to be a high-energy form of radiation.

On the heels of this discovery, French physicists Antoine Henri Becquerel (1852–1908) and Pierre Curie (1859–1906) and his wife, Marie Sklodowska Curie (1867–1934), discovered that a number of naturally occurring elements also exposed photographic plates that were sealed in light-tight containers. They had discovered that, in some cases, atomic nuclei are incapable of holding together and they spontaneously decay. This spontaneous process is called *radioactive decay*. Elements that undergo spontaneous nuclear decay—or *fission*—emit a variety of particles and rays, called alpha particles, beta particles, and gamma rays.

Alpha particles are made up of two protons and two neutrons and thus are positively charged. An alpha particle is identical to the nucleus of the helium atom. The addition of two electrons to an alpha particle gives a neutral helium atom. *Beta particles* are simply electrons, and the type of radioactive decay that produces them is also called *beta decay.* Finally, *gamma rays* are a form of high-energy electromagnetic radiation that is emitted by a nucleus when it undergoes fundamental change.

As we discussed in Chapter 15, the nucleus of an atom constitutes almost all its mass and almost none of its volume. The atom is mostly empty space. Neutrons and protons, referred to collectively as *nucleons,* have almost the same mass (the neutron has slightly more mass), and are about 2000 times more massive than the electron.

Isotopes

The *atomic number* of an element, that is, the *number of protons* in the nucleus of its atoms, determines its identity. For example, hydrogen has a single proton in its nucleus and is symbolized by ^{1}H. The addition of a neutron to the nucleus of a hydrogen atom gives *deuterium,* symbolized by ^{2}H. Deuterium is an *isotope* of hydrogen; that is, it is a form of hydrogen—since it has one proton in its nucleus—but it also contains a neutron. Any two or more species of an element that have the same atomic number but different numbers of neutrons are *isotopes.* Another isotope of hydrogen is *tritium,* with a proton and two neutrons in its nucleus, and is symbolized by ^{3}H. Whereas the hydrogen nucleus is stable (meaning that spontaneous decay is highly unlikely), the isotopes of hydrogen are radioactive. Water that contains these isotopes of hydrogen is sometimes called "heavy water," because the nuclei of these isotopes are more massive than the hydrogen nucleus.

PHYSICS IN THE REAL WORLD

There was a crisis at the Brookhaven National Laboratories (BNL) when it was announced in 1997 that groundwater had been contaminated with high levels of tritium. Largely in response to public outcry, in 1999 the apparent source of the contamination—the High Flux Beam Reactor (HFBR)—was permanently shut down.

The *atomic mass number* of an element is the total number of nucleons it contains. Thus, tritium has an atomic number of 1 and an atomic mass number of 3. Elements are sometimes written with a leading superscript and subscript referring to the atomic number and the atomic mass number of the element, respectively. Thus, carbon-14, a radioactive isotope of carbon used in the dating of living material can be written as $^{14}_{6}C$ or carbon-14. The most common isotope of carbon is carbon-12. Carbon-14 has two additional neutrons.

EXPLORATION 17.1

What would one call an isotope of helium that has one fewer neutron than normal? How about an isotope of helium with one more neutron?

The most common isotope of helium is helium-4, with 2 neutrons and 2 protons. If it had one less neutron, it would have only 3 nucleons, and would be helium-3. With one more neutron, it would be helium-5.

FORCES, FISSION, AND FUSION

There are four known forces in the universe. Until this chapter, we have discussed only two of them: *gravity* and the *electromagnetic force.* Gravity and electromagnetism are familiar to us because we interact with these forces on a daily basis. Gravity keeps us stuck to the ground, and holds the Moon in its orbit. The electromagnetic force governs electricity and magnetism, including the Coulomb forces of repulsion and attraction between charged particles, and the magnetic fields that moving charges generate.

As we discuss the nucleus we begin to realize the importance (and necessity) of two other forces, forces that have to do with the stability of the nuclei of atoms. These forces are no less crucial to our daily lives, but they are, perhaps, a little more difficult to appreciate.

As we pointed out earlier in the chapter, something keeps the nucleus together; not all nuclei decay like radioactive elements do. Some elements are stable for millions or billions of years. How can this be when the nuclei of all atoms other than hydrogen contain more than one proton? If the only force acting were the Coulomb repulsion between protons, then all nuclei would fly apart.

Strong Nuclear Force

In fact, there is another force at work in the nucleus; but first let's examine protons and neutrons more closely. These particles are themselves made up of even more fundamental particles called *quarks.* There are six different types of quarks; each quark type is called a *flavor* and has a whimsical name: *up, down, strange, charm, bottom,* and *top* (or *beauty* and *truth* to European physicists). The basic structure of a proton is two up quarks and one down quark, whereas a neutron is composed of two down quarks and one up quark. Such combinations of three quarks are called *hadrons,* meaning *heavy.* There are other hadrons more massive than the proton or neutron (made of heavier quarks), but they are unstable and very short-lived.

Counteracting the Coulomb force in the nucleus is the *strong nuclear force.* This force is the attractive force between protons that binds atomic nuclei together. The strong nuclear force is incredibly strong (far stronger than gravity or the electromagnetic force), but it also has an incredibly short range. It overwhelms all other forces if the two hadrons are within 10^{-15} m of each other, but if the two particles are farther apart, then Coulomb repulsion takes over, and the nuclei fly to pieces.

The necessity for this force begins to explain why some nuclei are highly radioactive, or decay easily. The nuclei of heavy elements, like uranium-235, for example, contain many nucleons. Uranium-235 contains 92 protons and 143 neutrons, to be exact. Its nucleus is so large that the electrical repulsion experienced by the protons is about as strong as the attraction of the strong nuclear force at the outer edges of the nucleus. Once a nucleus gets sufficiently large, it experiences radioactive decay on relatively short time scales. The amount of time it takes for half of a sample to decay is called its *half-life.* The half-lives of many substances are measured in years.

As mentioned previously, there are different types of radioactive decay. When an alpha particle is emitted, the atomic number decreases by 2 (2 protons are ejected) and the atomic mass number by 4 (2 neutrons are also ejected). A different type of decay (called *beta decay*) ejects a beta particle—an electron.

Weak Nuclear Force

The final force (the *weak nuclear force*) is involved in beta decay. During beta decay, a neutron decays to produce three other particles: a proton, an electron, and an antineutrino.

Neutron → Proton + Electron + Antineutrino

This decay occurs in the nucleus of an atom, and the nucleus ejects the beta particle (electron). The atomic number of the atom changes by 1, transforming it into another element (since an element is determined by its number of protons).

Some nuclei instead undergo beta plus decay in which a proton decays to become a neutron plus a *positron* (a particle with the same mass as an electron but with a positive charge; also called the antiparticle of an electron) and an electron-type *neutrino* (a particle with little or no mass and no charge).

The radioactive isotope of carbon called carbon-14 has a half-life of 5730 years. Carbon-14 decays via beta decay, meaning that the nucleus ejects an electron, and carbon-14 is changed into nitrogen-14. Because a tiny percentage of all carbon is this radioactive isotope of carbon, all living matter contains some carbon-14, and that level is maintained during its lifetime. Once an organism dies, the amount of carbon-14 it contains decreases by half every 5730 years.

EXPLORATION 17.2

What fraction of the original amount of carbon-14 would remain in wood that was chopped down 18,000 years ago?

We can simplify this problem by rounding the half-life slightly to 6000 years. That means that half of the original amount will be left after 6000 years, one-fourth will be left after 12,000 years, and one-eighth will be left after 18,000 years.

Nuclear fusion is in many ways the inverse of the process of fission. *Fission* is the breaking apart of atomic nuclei. *Fusion* is the generation of heavier nuclei from lighter ones. As you might imagine, nuclei resist being pushed together (because of the repulsion of their like charges), so conditions must be unusual for fusion to be able to occur. Temperature and density must be extremely high, and these conditions exist naturally in the core of the largest fusion reactors that we know of: stars.

Nuclear Fusion in the Sun

The core of the Sun, or any other star, starts out as mostly hydrogen. The temperatures of stellar cores are at least 10 million kelvin and under intense pressures, and under these conditions, protons can collide with sufficient energy to overcome the barrier of their Coulomb repulsion and fuse.

The process of hydrogen fusion in the cores of stars has four major products that we have discussed: helium, energy (in the form of gamma rays), positrons, and neutrinos.

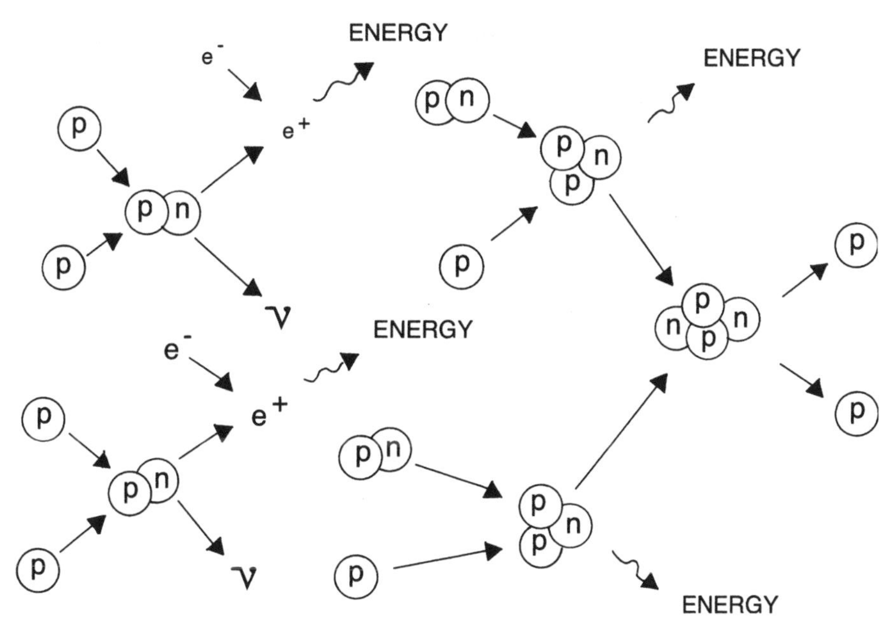

Figure 17.1 The Fusion of Helium That Occurs in the Cores of Stars

When two protons collide, one of them becomes a neutron, which combines with the other proton to generate a deuterium nucleus, or deuteron. A positron and a neutrino are also produced. The positron collides with an electron to produce gamma rays. The deuteron can then combine with another proton to produce helium-3, which is an isotope of the helium atom that has a nucleus containing two protons and a neutron. This fusion reaction generates energy in the form of a gamma ray. Finally, two helium-3 nuclei fuse to form helium-4 and two hydrogen nuclei (protons). This entire process is known as the *p-p chain* and is shown schematically in Figure 17.1.

PHYSICS IN THE REAL WORLD

Because the neutrinos produced in stellar reactions are nearly massless and chargeless, they do not interact much with other matter and tend to flow nearly unimpeded through the universe. The neutrons produced in the core of the Sun in the reactions described, for example, stream out of the Sun and through Earth. Billions of them sweep through your body every second.

ELEMENTARY PARTICLES AND FORCES: A SUMMARY

The "indivisible" atom of the ancients apparently can be broken up in a variety of ways, and we take a pause here to take stock of the particles and forces that we have named. The atom consists of electrons, protons, and neutrons. The (1) *electromagnetic force* holds the electrons and the protons together, and the (2) *strong nuclear force* holds protons together in the nucleus of the atom. The (3) *weak nuclear force* is related to the stability of the neutron itself and its potential to decay into a proton. On the scale of atoms and nuclei, the (4) *gravitational force* is

PHYSICS IN THE REAL WORLD

The electromagnetic force is responsible for many concrete forces, such as the frictional forces between objects, and the force that resists your mass pushing down on the surface of Earth. It is the repulsion between the electron shells surrounding the molecules in your body and the electron shells surrounding molecules in the ground that keep you (and everything around you) from sinking into Earth, which is, after all, made of atoms that are mostly empty space.

insignificant but dominates on scales larger than that of Earth.

Table 17.1 lists the four fundamental forces, their relative strengths, and the scales over which they are significant.

THE BIG BANG

We end our journey with a brief summary of the origin of the universe that we inhabit, as we understand it at the dawn of the 21st century. In the 1920s astronomer Edwin Hubble (1889–1953) observed that galaxies around our own were moving away from us, and that more distant galaxies were moving away more rapidly. This *redshift* in galaxy velocities seemed to imply not that we are in a special place (with everything moving away from us) but that all galaxies in the universe are moving away from one another. Any other inhabitant of any other galaxy should observe the same motion.

Force	Relative Strength	Range (m)
Strong nuclear	1	10^{-15}
Electromagnetic	0.01	infinite
Weak nuclear	10^{-7}	10^{-15}
Gravity	10^{-38}	infinite

Table 17.1

At about the same time, Einstein was working on his theory of general relativity, but he encountered a problem. The distortion of space–time owing to gravity pulled his universe together, but what kept it from contracting completely? How did the universe exist? Einstein proposed a gravitational constant that acted almost like antigravity and supported his theoretical universe.

When Einstein learned of the observation of the recession of galaxies, he called his proposed constant his biggest blunder. The observed expansion alone could apparently support the universe from gravitational collapse.

PHYSICS IN THE REAL WORLD

It turns out that physicists at the dawn of the 21st century have had to revisit Einstein's "blunder" to help explain why observations of galaxies at very high redshift—meaning that they are very distant—indicate that the expansion of the universe is accelerating; that is, the universe is expanding more rapidly now than in the past.

One way to explain this expansion is that the universe is imbued with a "dark energy" that behaves in a way that can be described by Einstein's gravitational constant. It may not have been a blunder after all. Figure 17.2 shows an image of one of the most distant supernovas ever detected.

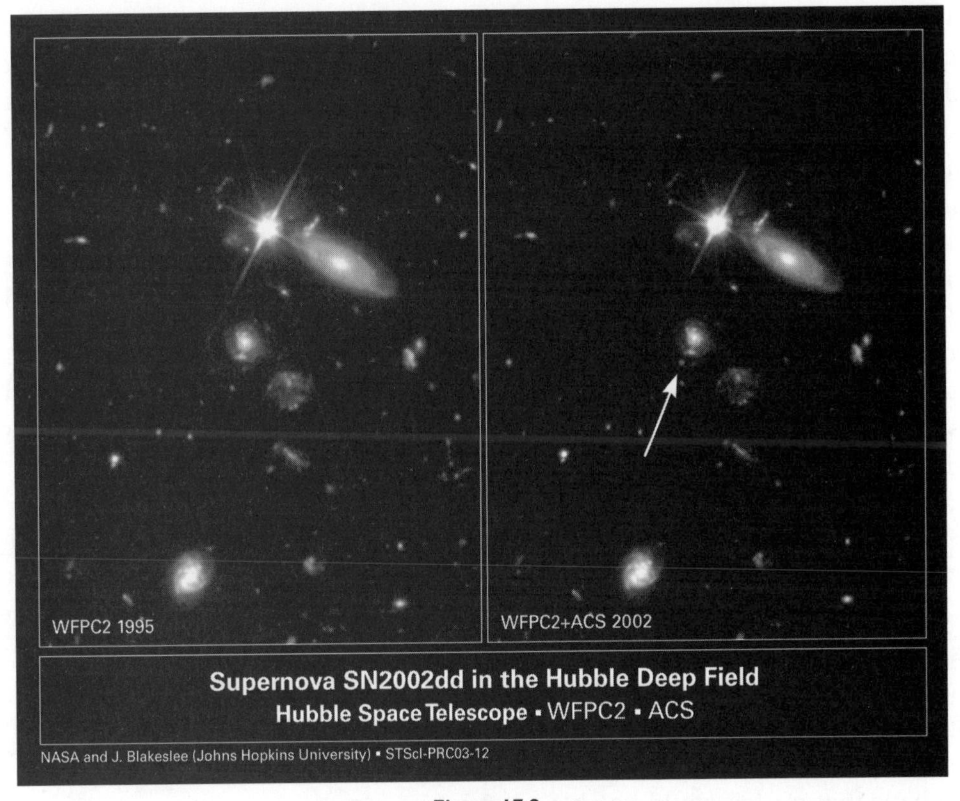

Figure 17.2
The arrow indicates the light from a very distant supernova exploding, at a distance estimated to be 8 billion light years. Distant supernovae like this one suggest that the acceleration of the universe is expanding. Photo courtesy of NASA and J. Blakeslee (JHU).

The expansion of the universe was interpreted as evidence of an initial explosion that by the middle of the twentieth century had been dubbed the *Big Bang*. The story of the early universe is one of cooling and expansion.

At about 1/100 of a second after the Big Bang, the entire universe was incredibly hot (10^{11} K) and filled with the elementary particles that we have been discussing in this chapter, the building blocks of atoms: electrons, positrons, neutrinos, and photons of light. There were a lot more light things (electrons) than heavy things (neutrons and protons) at the time, on the order of about a billion to one.

Einstein would have liked this early time in the universe, since the equivalence between energy and matter was obvious. With the entire universe hotter than the core of a star, energy was freely being converted to matter, and matter to energy. Electrons and positrons *annihilated one another* to create energy carried away by energetic photons, and electrons and positrons were generated from the energy carried by other photons.

At only 1 s after the Big Bang, the temperature had cooled to 10^{10} K, still too hot for neutrons and protons to be held together in nuclei by the strong nuclear force.

As the universe continued to cool, soon there was not enough energy to create new electrons and positrons from the photons that filled space. As a result, most of these light positively and negatively charged particles annihilated one another, and no new ones took their place. Finally, less than 3 min into the life of the universe, it was cool enough for helium-4 nuclei to form via the fusion of hydrogen nuclei and not be destroyed.

After the first 3 min, the universe had cooled to about 10^9 K, and most of the electrons and positrons had annihilated one another (generating photons). The universe consisted of photons, neutrinos, and anti-neutrinos, and a relatively small number of nucleons (neutrons and protons).

Because the universe was at this point as hot and dense as the core of a star undergoing nuclear fusion, the universe was still far too hot for electrons to link with nuclei (via the electromagnetic force) to form stable atoms. It would not be until almost 300,000 years after the big bang that nuclei would be able to hold on to electrons. This meant that until 300,000 years into the birth of the universe, it was filled with a "fog" of electrons. Once the universe cooled enough for nuclei to hold onto electrons, the fog cleared, and the photons that had been trapped in the fog were free to expand into the universe.

The photons that were generated at that moment are now referred to as the *cosmic microwave background* (CMB). The CMB microwaves are these photons that escaped at that moment and were redshifted to radio wavelengths by the expansion of the universe itself.

From this point on, the universe continued simply to cool and coalesce and expand, eventually forming the stars and galaxies of the observable universe. The importance of gravity now became paramount, as the average separation between particles was measured in units much larger than 10^{-15} m. The strong and weak nuclear forces continued to govern the lives of the nuclei of atoms, but not of the entire universe.

PHYSICS IN THE REAL WORLD

Have you ever seen the movie Poltergeist, or just stared at a television screen that had no signal? That gray speckled pattern is sometimes called snow. The snow is just due to radio "noise" picked up by the receiver of your television set. It has been estimated that about 1 percent of that snow is due to the cosmic microwave background. The next time you awake at 3 in the morning in front of snow on the television screen, remember that you are looking at an echo of the birth of the universe.

This is the universe that we now inhabit, orbiting one planet around a single star in a sea of stars, in the outer reaches of a typical galaxy in a sea of galaxies.

What will be the eventual fate of the universe? Will it expand forever? We may not be sure of all the details yet, but we can be sure that the laws of physics will lead us to a deeper understanding.

PROBLEMS

17.1 What is the difference between fission and fusion? Which of these two is the energy-generation mechanism of the Sun?

17.2 How is an alpha particle different from a proton? How is it similar to a proton?

What is the difference between helium-3 and helium-4?

17.3 Why is the strong nuclear force essential to the stability of an atom? What would happen to the universe if there were no force like the strong nuclear force?

17.4 Explain why, in terms of the strong nuclear force and the electromagnetic force, larger nuclei are more likely to undergo radioactive decay.

17.5 Some radioactive waste has a half-life of 10,000 years. If there is a 1000-kg pile of radioactive waste on Earth now, how many kilograms will be left (still radioactive) after 40,000 years?

17.6 A sample of wood is found to have $^{1}/_{32}$ of the carbon-14 that would be expected in living matter. How old is the wood?

17.7 Which of the four forces was most important in shaping the orbits of the planets in the solar system?

17.8 How much energy is needed to create an electron and a positron? (*Hint:* Use Einstein's mass–energy equivalent.)

17.9 How much energy is needed to create a proton and an antiproton? (*Hint:* Protons and their antiparticles are about 2000 times more massive than an electron.)

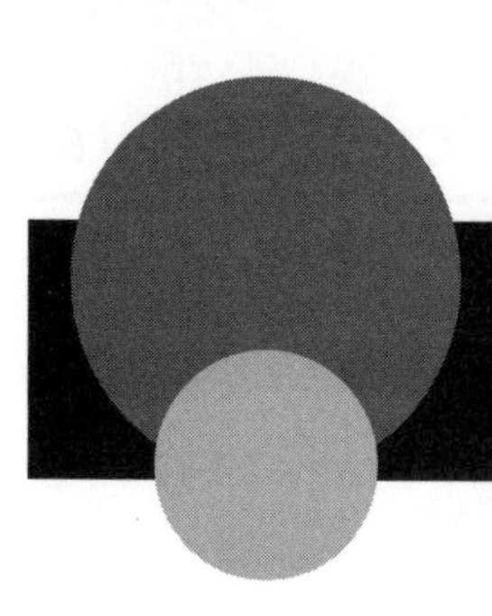

APPENDIX A FORMULAS AND RELATIONS

Measurement

Density of a substance: $D = \frac{M}{V}$, where D is the density, M is the mass of a sample of the material, and V is the volume of that sample

Area of a circle: $A = \pi r^2$, where r is the radius of the circle

Surface area of a sphere: $A = 4\pi r^2$, where r is the radius of the sphere

Volume of a sphere: $V = {}^4/_3\pi r^3$

Volume of a cylinder: $V = \pi r^2 h$, where r is the radius of the circular base and h is the height of the cylinder

Force

Newton's second law: $F = ma$, where F is the force acting on a body, a is the acceleration of the body's motion, and m is the mass of the body

Law of gravitation: $F = Gm_1m_2/d^2$, where F is the force of attraction between two masses, G is the universal gravitation constant, m_1 and m_2 are the two masses, and d is the distance between them

Linear Motion

Average speed of motion: $v = \frac{d}{t}$, where v is the average speed, d is the distance covered, and t is the elapsed time

Position after time t at constant velocity: $x = x_0 + vt$, where x_0 is the initial position, v is the velocity, and t is time

Acceleration: $a = \frac{v}{t}$, where v is the change in speed, and t is the time required to produce that change

Velocity after time t at constant acceleration: $v = v_0 + at$, where x_0 is the initial position, a is the acceleration, and t is time

Position after time t in freefall: $x = x_0 + v_0t + \frac{1}{2}at^2$, where x_0 is the initial position, v_0 is the initial velocity, and g is the acceleration of gravity (9.8 m/s^2)

Momentum: $p = mv$, where p is the momentum, m is the mass of the body, and v is its velocity

Circular Motion

Angular velocity: $\omega = \Delta\theta/\Delta t$, where θ is angular position and t is time

Angular acceleration: $\alpha = \Delta\omega/\Delta t$, where ω is angular velocity and t is time

Angular velocity: $\omega = v/r$, where v is the linear velocity and r is the radius of curvature of the circular motion

Torque, or turning effect, of a force about a given pivot point: $\tau = rF$, where τ is the torque, F is the amount of the force, and r is the perpendicular distance from the pivot to the line of the force

Angular momentum: $L = mvr$, where m is the mass of an object moving at velocity v in a circle of radius r

Work and Energy

Work done by a force: $W = Fd$, where W is the amount of work done, F is the magnitude of the force, and d is the distance moved in the direction of the force

Kinetic energy of a moving body: KE = $\frac{1}{2}mv^2$, where KE is the kinetic energy, m is the mass, and v is the speed of the body; kinetic energy is measured in joules

Gravitational potential energy: GPE = mgh, where m is the mass of the object, g is the acceleration of gravity, and h is the height

Power: $P = \frac{W}{t}$, where P is the average power expended, W is the amount of work done, and t is the time required to do it

Heat and Thermal Energy

Celsius and Fahrenheit temperatures: Readings on the two scales are related by $F = \frac{9}{5}C + 32$, where C is any temperature on the Celsius scale and F is the corresponding temperature on the Fahrenheit scale

Heat required to change the temperature of a body: $Q = mc\,\Delta T$, where Q is the quantity of heat absorbed or given off, c is the specific heat of the material, m is the mass of the body, and ΔT is its temperature change

Length change with temperature: $\Delta L/L = \alpha\Delta T$, where ΔL is the change in length, L is the length of the object, a is the coefficient of linear expansion, and ΔT is the temperature change

Pressure

Boyle's law: If the temperature of a gas remains constant, the pressure and volume are inversely proportional: $\frac{V_1}{V_2} = \frac{p_2}{p_1}$, where p_1 and V_1 are the original pressure and volume and p_2 and V_2 are the resultant pressure and volume, respectively

Pressure: $P = F/A$, where F is the exerted force, and A is the area over which the force acts

Pressure at depth h: $P = Dgh$, where D is the density of the liquid, g is the acceleration of gravity, and h is the depth in the liquid

Archimedes' law: Buoyant force on a body immersed in a liquid = Weight of liquid displaced by the body

Wave Motion

Frequency and period: $f = 1/T$, where f is the frequency of a wave, and T is its period

Velocity of a wave: $v = \lambda f$, where λ is the wavelength and f is the frequency

Velocity of a sound wave: $v = (331 + 0.6T)$ m/s, where T is the air temperature in degrees Celsius

Doppler effect for a moving source: $f_0 = (v/v - v_s)f_s$, where f_0 is the frequency detected by the observer, f_s is the frequency emitted by the source, v is the velocity of the wave, and v_s is the velocity of the source

Light

Velocity of light: $c = f\lambda$, where f is the frequency of the radiation, λ is its wavelength, and c is the speed of light (3×10^8 m/s)

Doppler shift for light: $\Delta\lambda/\lambda = v/c$

Illumination produced by a small light source on a surface held perpendicular to the rays: $E = C/d^2$, where E is the illumination, C is the intensity of the source, and d is its distance from the illuminated surface

Index of refraction: $n = c/v$, where n is the index of refraction of a material in which the speed of light is v, and c is the speed of light in a vacuum

Location of image formed by a converging lens $\frac{1}{p} + \frac{1}{q} = \frac{1}{f}$, where p is the distance of the object from the lens, q is the distance of the image from the lens, and f is the focal length of the lens

Size of the image: $h_i/h_o = q/p$, where h_i is the height of the image, h_o is the height of the object, q is the image distance, and p is the object distance

Electricity

Electric force: $F_C = kq_1q_2/r^2$, where q_1 and q_2 are the two charges, r is the separation between them, and k is the Coulomb constant

Electric field strength: $E = F/q$, where E is the electric field strength, F is the coulomb force, and q is the charge of the particle

Potential difference: $PD = \Delta EPE/q$, where ΔEPE is the change in electric potential energy, and q is the charge. The unit of potential difference is the volt (V), where 1 V = 1 J/C.

Strength of an electric current: $I = Q/t$, where I is the current strength, Q is the total quantity of charge passing any point in the conductor, and t is the time during which it passes

Ohm's law: $I = V/R$, where I is the strength of the current flowing in a conductor, V is the potential difference applied to its ends, and R is its resistance

Electric power: $P = IV$, where P is electric power, I is current, and V is voltage. Power is measured in watts (W).

Resistors in series: $R = R_1 + R_2 + R_3 + \cdots + R_n$

Resistors in parallel: $\frac{1}{R} = \frac{1}{R_1} + \frac{1}{R_2} + \frac{1}{R_3} + \text{Á} + \frac{1}{R_n}$

Capacitance: $C = q/PD$, where q is the charge, and PD is the potential difference between two plates. Capacitance is measured in farads (1 F = 1 C/V).

Magnetic field around a current: $B \propto I/2\pi r$, where I is the current and r is the distance from the wire

Force on a moving charge: $F = qvB$, where q is the charge, v is the velocity of the charge, and B is the strength of the magnetic field; the direction of the force is determined by the right-hand rule, as described in the text

Transformer voltage and current: $\frac{V_s}{V_p} = \frac{n_s}{n_p}$, where V_p is the voltage in the primary coil, V_s is the voltage in the secondary coil, and n_p and n_s, are, respectively, the numbers of turns in each $\frac{I_s}{I_p} = \frac{n_p}{n_s}$ is the corresponding expression for the current

Quantum Mechanics, the Atom, and Relativity

Energy of a photon: $E = hf$, where h is Planck's constant (6.63×10^{-34} J•s) and f is the photon frequency; can also be expressed as $E=hc/\lambda$

de Broglie wavelength: $\lambda = h/p$, where h is Planck's constant and p is the momentum

Time dilation: $t = t_0/\sqrt{[1 - v^2/c^2]}$, where t is the time, t_0 is the relative time, v is the relative velocity, and c is the speed of light

Relativistic mass: $m = m_0/\sqrt{[1 - v^2/c^2]}$, where m is the original mass, m_0 is the relative mass, v is the relative velocity, and c is the speed of light

Lorentz factor: $\gamma = 1/\sqrt{[1 - v^2/c^2]}$, where v is the relative velocity and c is the speed of light

Mass–energy equivalence: $E = mc^2$, where E is the energy in ergs, equivalent to a mass m in grams, and c is the speed of light in centimeters per second

APPENDIX B MOTION IN TWO DIMENSIONS

When objects move through space in the real world, they rarely move in one dimension, as discussed in Chapter 3. Real objects (birds, airplanes, balls) move through three-dimensional (3-D) space. It is possible to model motions in three dimensions but more complicated than we need to tackle in this appendix. Here we extend the discussion of motion in one dimension to motion in two dimensions, sometimes called motion in the *x-y* plane. This type of analysis is particularly useful for analyzing the motions of objects through space under the influence of gravitational forces.

In Chapter 3 we looked at the example of accelerated motion in one dimension (free fall). Basically, an object released from rest or at some velocity up or down will fall under the influence of gravity, and one can easily calculate the position and velocity of the object as a function of time.

A more realistic example involves an object being thrown at some angle to the horizon (see Figure B-1). In this image, a girl throws a ball that follows a path or trajectory through two-dimensional (2-D) space. The shape of the trajectory is called a *parabola,* and it is the path followed by an object launched at an angle to the horizon, under a gravitational force, neglecting the force of air resistance (air friction). If there were no gravitational force, the ball would move off at a constant velocity in a straight line (see Figure B-1).

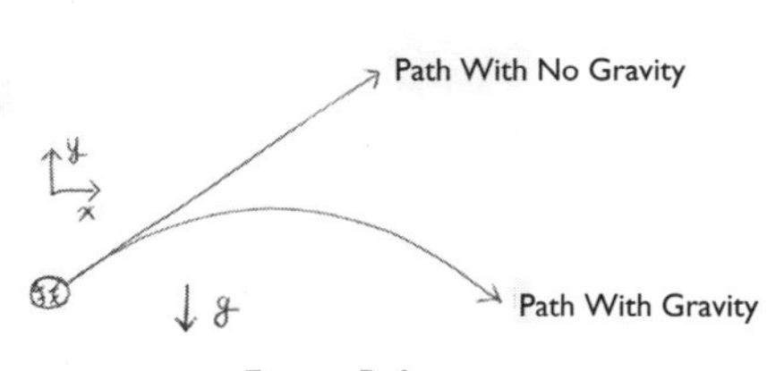

Figure B-1

Suppose that the ball is thrown at an angle of 45° to the horizon (see Figure B-2), and the initial velocity is 10 m/s. It is much easier to solve a problem like this if we break the velocity vector into two *component vectors*. The component vectors are perpendicular to each other, one in the *x*-direction and one in the *y*-direction. Using some basic rules of trigonometry, we can determine the two component velocities:

Velocity in the *x*-direction:

$$v_x = v \cos \theta = (10 \text{ m/s}) \cos(45°) = 7.1 \text{ m/s}$$

where θ is the angle to the horizon (see Figure B-2).

Velocity in the *y*-direction:

$$v_y = v \sin \theta = (10 \text{ m/s}) \sin(45°) = 7.1 \text{ m/s}$$

where θ is again the angle to the horizon (see Figure B-2).

The *x*- and *y*-velocities are equal in this case, because the angle is 45° to the horizon. If the angle were greater than 45°, the velocity would be greater in the *y*-direction; if the angle were less than 45°, the velocity would be greater in the *x*-direction.

What happens to the ball as it moves through space?

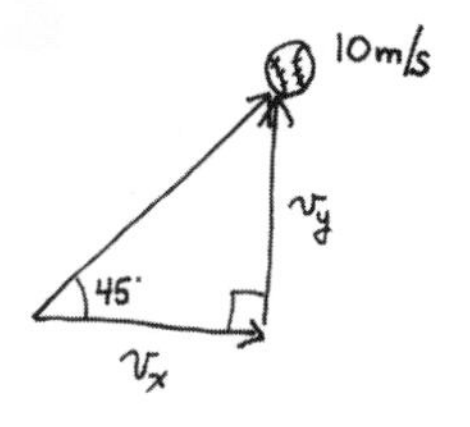

Figure B-2

There are no forces acting in the x-direction (if we neglect air resistance), so the ball moves at constant velocity in the x-direction.

In the y-direction, however, there is a force acting on the ball, the force of gravity. We can use the velocity equation from Chapter 3 to determine how long it will take for the ball to stop moving upward (as we know it will) and start moving downward. We use the equation

$$v = v_0 + gt$$

where v is the final velocity in the y-direction (0 m/s; it stops instantaneously), v_0 is the initial velocity in the y-direction (7.1 m/s), g is the acceleration of gravity, and t is the time.

Solving for t, we find that

$$t = (v - v_0)/g$$

$$t = (0 - 7.1 \text{ m/s})/(-9.8 \text{ m/s}^2)$$

where we use a negative value for g, since the velocity is positive upward and acceleration of gravity is positive downward (see Figure B-1).

As a result, we find that after 0.72 s the ball begins its journey back to Earth. Because the trajectory is completely symmetric (check this if you like), the return journey takes another 0.72 s, for a total of 1.4 s.

How far away from the girl will the ball land?

Moving at 7.1 m/s (horizontally) for 1.4 s, the ball will travel

$$D = v \times t$$

$$D = (7.1 \text{ m/s})(1.4 \text{ s}) = 9.94 \text{ m}$$

As you might imagine, the ball will land at varying distances from the girl depending on two factors, the velocity of the ball and the angle to the horizon.

APPENDIX C SOLUTIONS TO PROBLEMS

Chapter 2

2.1 Using a height of 6 ft 3 in: $(6 \times 12) + 3 = 75$ in. $\times$ 2.54 cm/in. = 190.5 cm or 1.9 m

2.2 SA $= 4\pi r^2 = 4(3.14)(6400 \text{ km})^2 = 5.15 \times 10^8$ km^2; $\frac{2}{3}(5.15 \times 10^8 \text{ km}^2) = 3.43 \times 10^8$ km^2

2.3 (a) 2.99792458×10^8 m/s; (b) 8.64×10^4 s; (c) 9.192631770×10^9 oscillations; (d) 1.5×10^{-5} cm

2.4 Volume $= \frac{4}{3}\pi r^3 = (4.2)(10 \text{ cm})^3 = 4.2 \times 10^3$ cm^3; $D = M/V = (10{,}000 \text{ g}) \div (4.2 \times 10^3 \text{ cm}^3) = 2.4$ g/cm^3; an alloy of aluminum

2.5 1×10^6; 0.1

2.6 60 s/min $\times$ 60 min/h $\times$ 24 h/day = 86,400 s/day; 86,400 s/day $\times$ 30 day/month = 2,592,000 s; 3.16×10^7 s

2.7 There are approximately 300 million Americans. If there is one car per person, then 300 million cars $\times$ 12,000 miles per year (as per most car warranties) $\times$ 1 gal $\div$ 20 mi = 180 billion gallons of gas

2.8 Average family of 4 eats 2 fast-food meals each week. Each meal costs \$5 per person. Therefore, 4 persons $\times$ 2 meals/person $\times$ \$5/meal $\times$ 300 million Americans = \$12 billion

2.9 300 million persons $\times$ 1 family/4 persons $\times$ 2 cars/family = 150 million cars

Chapter 3

3.1 Speed = distance/time = 250 mi/5 h = 50 mi/h

3.2 Time = distance/velocity = (1 mi)/(10 mi/h) = 0.1 h, or 6 min

3.3 Use the Pythagorean theorem: Speed $= \sqrt{(250 \text{ mph})^2 + (60 \text{ mph})^2} =$ 257 mi/h, to the southeast

3.4 $0 = 0$ m/s + (9.8 m/s^2) (10 m)(60 s/min) $= 5.9 \times 10^3$ m/s, or 5.9 km/s; (5.9 km/s) (3600 s/h) $= 2.1 \times 10^4$ km/h

3.5 = 50 m + (0 m/s)t + (−9.8 m/s)t^2; $t = \sqrt{(50 \text{ m})/(9.8 \text{ m/s}^2)} = 2.3$ s

3.6 = 1 m + (0 m/s)t + (−9.8 m/s^2)t^2; $t = \sqrt{(1 \text{ m})/(9.8 \text{ m/s}^2)} = 0.31$ s; both quarters should hit the ground at the same time (see Appendix B)

Chapter 4

4.1 magnitude, direction

4.2 inertia

4.3 mass, force, acceleration

4.4 212 N; his other arm

4.5 4900 N; no, as long as the truck is in the middle of the bridge, and we can neglect the mass of the bridge, and the bridge does not flex.

4.6 980 N; 980 N; 1080 N; That will be its acceleration in freefall; no force other than gravity is required, which is −980 N.

Chapter 5

5.1 Work = (200 N)(10 m) = 2000 N•m = 2000 J = 2 kJ

5.2 PE = (500 kg)(9.8 m/s^2)(10 m) $= 4.9 \times 10^4$ J; (3 kg)(9.8 m/s^2)(200 m) $= 5.9 \times 10^4$ J (the bowling ball); the work required is equal to the PE of each object

5.3 KE $= \frac{1}{2}(1.67 \times 10^{-23} \text{ kg})\,(0.03 \times 2.99 \times 10^8 \text{ m/s})^2 = 6.7 \times 10^{-10}$ J; $(6.7 \times 10^{-10} \text{ J})\,(1 \text{ eV}/1.6 \times 10^{-19} \text{ J}) = 4.2 \times 10^9$ eV

5.4 PE + KE (bridge) = PE + KE (river); 50 m $\times$ 9.8 m/s^2 $\times$ 0.1 kg + 0 = 0 + $\frac{1}{2}$(0.1 kg)v^2; $v = \sqrt{980 \text{ m}^2/\text{s}^2} = 31$ m/s

5.5 Assuming all her potential energy is converted to kinetic energy, PE + KE (top) = PE + KE (bottom); $(2\text{ m})(9.8\text{ m/s}^2)(25\text{ kg}) + 0 = 0 + \frac{1}{2}(25\text{ kg})\,v^2$; $v = \sqrt{39.2\text{ m}^2/\text{s}^2} = 6.3\text{ m/s}$

5.6 Work = force × distance = $(10{,}000\text{ kg})(9.8\text{ m/s}^2)(0.5\text{ m}) = 4.9 \times 10^4\text{ J}$; Power = work/time = $(4.9 \times 10^4\text{ J})/(30\text{ min})(60\text{ s/min}) = 27.2\text{ J/s} = 27.2\text{ W}$

Chapter 6

6.1 180°; π rad; $20 \times 360° = 7200°$; $20 \times 2\pi = 40\pi$ rad

6.2 $33\frac{1}{3}$ rev; $(33.33\text{ rev/min})(2\pi\text{ rad/rev})(1\text{ min}/60\text{ s}) = 3.5\text{ rad/s}$; $(45\text{ rev/min})(2\pi\text{ rad/rev})(1\text{ min}/60\text{ s}) = 4.7\text{ rad/s}$

6.3 2π rad/5 s = 0 + angular acceleration (5 s); ang. accel. = 0.25 rad/s^2

6.4 The torque increases with distance from the pivot point (r), so that less force is required at a greater distance (to exert the same torque). Torque = force × 0.3 m

6.5 As they move to the center of the merry-go-round, they are concentrating their mass closer to the axis of rotation. To conserve angular momentum, the merry-go-round rotates faster. This same effect causes an ice skater to spin faster when she pulls in her arms while spinning.

6.6 $\omega = 2\pi\text{ rad}/300\text{ s} = 0.02\text{ rad/s}$; $v_t = r\omega = (1\text{m})(0.02\text{ rad/s}) = 0.02\text{ m/s}$; $v_t = r\omega = (3\text{m})(0.02\text{ rad/s}) = 0.06\text{ m/s}$; angular velocity is unchanged

6.7 Fg (Venus)/Fg (Earth) = $[Gm(\text{V})m(\text{S})/r(\text{V}-\text{S})^2]/[Gm(\text{E})m(\text{S})/r(\text{E}-\text{S})^2]$; most terms cancel (since $m(\text{V}) = m(\text{E})$, and the G and $m(\text{S})$ terms cancel). This leaves $Fg(\text{V})/Fg(\text{E}) = [r(\text{E}-\text{S})/r(\text{V}-\text{S})]^2$; $(4/3)^2 = 1.8$ times S; Venus experiences a force that is 1.8 times larger than that experienced by Earth.

Chapter 7

7.1 Neutrons and protons are the most massive. Neutrons are slightly more massive than protons; protons, +1; electrons, −1; neutrons, 0

7.2 An atom comprises a nucleus (containing neutrons and protons) and electrons. It is an irreducible form of a chemical element. A molecule is a collection of atoms, held together by chemical bonds. An element is a collection of the same atoms. Elements have distinct chemical properties and numbers of protons in their nucleus; compounds are substances that consist of more than one element; the elements in our bodies come from stars, mostly.

7.3 $(75\text{ kg})(1\text{ atom}/1.67 \times 10^{-27}\text{ kg}) = 4.5 \times 10^{28}$ atoms; far more than the number of stars in the Galaxy

7.4 Density = $(75\text{ kg})/(0.064\text{ m}^3) = 1.2 \times 10^3\text{ kg/m}^3$; she will sink.

7.5 A lead ball will float in mercury; a gold ball will sink; compare their relative densities.

7.6 Measure the mass of 50 pennies; then use displacement of water in a graduated cylinder to determine their exact volume. Density = mass/volume. Compare the derived density with that of copper.

7.7 Pressure decreases with altitude; gravity holds most of the atmosphere close to the surface of Earth, and the smaller number of molecules at high altitude exert less pressure.

7.8 Assuming she is completely submerged: Buoyant force (water) = weight of water displaced = $(0.064\text{ m}^3)(1000\text{ kg/m}^3)(9.8\text{ m/s}^2) = 630\text{ N}$; if she were in mercury, the force would be $(0.064\text{ m}^3)(13{,}600\text{ kg/m}^3)(9.8\text{ m/s}^2) = 8500\text{ N}$; note that if she were submerged, she would accelerate up, since the buoyant force is greater than her weight (in newtons).

Chapter 8

8.1 100°F = $9/5T + 32$; $T = 38°\text{C}$; $T(\text{K}) = 311\text{ K}$

8.2 A change of 2°C is the same as a change of 2 K; T (F) = 9/5(2) + 32; 1.8°F

8.3 Because it contains a lot of internal heat and cools slowly; because the insulation would not allow heat to easily pass from the water to your toes.

8.4 Q (Al) = $(1\text{ kg})(920\text{ J/kg}\cdot°\text{C})(20°\text{C}) = 1.8 \times 10^4\text{ J}$; $Q\ (H_2O) = (0.5\text{ kg})(4186\text{ J/kg}\cdot°\text{C})(5°\text{C})$

$= 1.1 \times 10^4$ J; the aluminum requires more energy.

8.5 Energy = $(3.33 \times 10^5$ J/kg)(1 kg) $= 3.33 \times 10^5$ J

8.6 Allowing for temperature variations between −20°C and 40°C: $\Delta L = L\alpha\Delta T$ = (100 cm)(0.000012)(60°C) = 0.07 cm; a few millimeters should be sufficient.

8.7 Living things appear to maintain a state of order; but if we consider living things and all the waste they generate, then that whole system is increasing its entropy (disorder).

8.8 As you accelerate, the density of air increases at the back of the car (as more molecules accumulate there owing to their inertia). As a result, the balloon feels the pressure difference and moves forward.

8.9 If the house is airtight, dangerous pressure differences can arise inside and outside the house. If windows are opened, inside and outside pressure can equalize and protect the house.

8.10 Conduction: a spoon in soup heats up; convection: hot air rises from a flame to fill a balloon; radiation: your skin heats up as you sit in direct sunlight, absorbing photons generated in the core of the Sun.

Chapter 9

9.1 Wavelength = 3 m; period = 1 s; velocity = 3 m/s

9.2 The waves could not constructively interfere, they would just move past one another. This is because there is no way for the human wave to double its "amplitude." The amplitude is defined by the standing height of the crowd.

9.3 See figure C-9.3

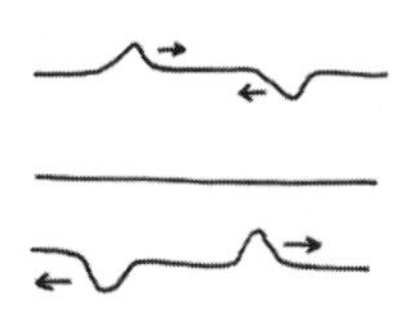

Figure C-9.3

9.4 That he is wasting your money. Humans cannot hear frequencies that high (beyond 20,000 Hz).

9.5 Speed of sound = 331 m/s + 0.6(100) = 391 m/s

9.6 Because at higher temperatures molecular collisions are more frequent, and sound is transferred through these sorts of collisions (longitudinal waves).

9.7 Observed frequency = (349 m/s)/(349 m/s + 10 m/s) 256 Hz = 249 Hz; lower frequency; lower pitch

9.8 v = 331 + 0.6(30) = (349 m/s)(1 km)/(1000 m)(3600 s/h) = 1300 km/h

9.9 As the temperature drops, the speed of sound drops, meaning that it should be easier to produce a sonic boom at high altitude.

9.10 Whisper is about 10 dB; ordinary conversation is about 50 dB. Difference is 40 dB, so ordinary conversation is about 10^4, or 10,000 times louder.

Chapter 10

10.1 Olive oil: 1.47 = c/v, so that $v = c/1.47 = 2.0 \times 10^8$ m/s; diamond: 2.42 = c/v, so that $v = c/2.42 = 1.2 \times 10^8$ m/s

10.2 Time = (50 mi)/(183,310 mi/h)(3600 s/h) = 0.98 s

10.3 10 ft-c = C/(30 ft)^2; C = 9000 candles

10.4 See figure C-10.4.

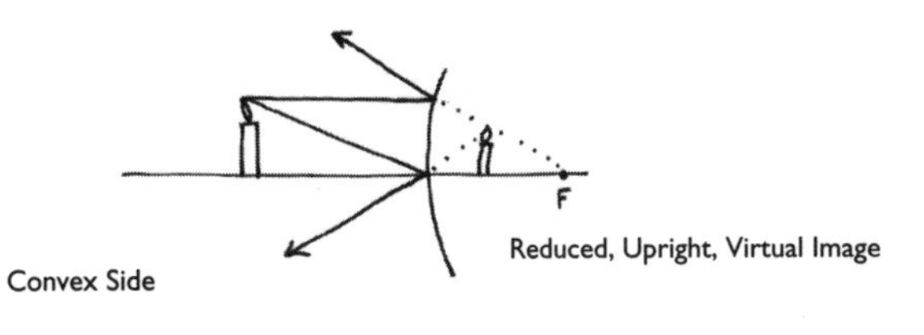

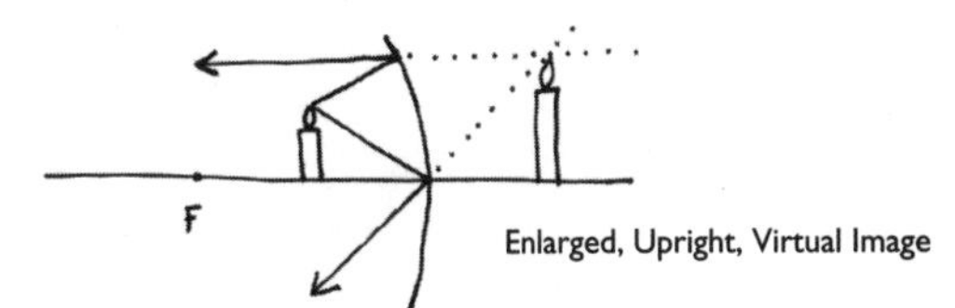

Figure C-10.4

10.5 See Figure 10.5 on page 89.

10.6 You will form a virtual, upright image. The image will be larger than the object. This is a magnifying glass!

10.7 It is a plane mirror; you can tell because objects in that mirror are not closer than they appear. The driver's side rear-view mirror is also a plane mirror.

10.8 Mirrors have less mass and are easier to move around than lenses, especially when they are very large. The index of refraction is different for different wavelengths, which can cause chromatic (color) aberration.

Chapter 11

11.1 Because you can see refraction from water particles above and below you. On the ground, you can see only the refraction occurring above you.

11.2 Frequency = c/wavelength = $(3 \times 10^8$ m/s)/(s/656.3 $\times 10^{29}$ m) = 4.6×10^{14} Hz

11.3 Frequency = c/wavelength = $(3 \times 10^8$ m/s)/(0.21 m) = 1.4×10^9, or 1.4 GHz

11.4 Yellow; black

11.5 Change in wavelength/wavelength = v/c; v = 3×10^8 m/s (200 nm/700 nm) = 8.6×10^7 m/s, or about 30 percent of the speed of light!

11.6 Reflection gratings are cheap to manufacture, since they are a thin coating on an inexpensive material (plastic).

11.7 Emission line spectrum; they contain hot gas.

11.8 Blackbody spectrum; it is a hot solid.

Chapter 12

12.1 No, because metal is a much better conductor than plaster or gypsum wallboard

12.2 $F_g = (6.67 \times 10^{-11}$ N m^2/kg^2)(9.11 $\times 10^{-31}$ kg)(1.67 $\times 10^{-27}$ kg)/(5.3 $\times 10^{-11}$ m)^2 = 3.6×10^{-47} N, far smaller than the Coulomb force

12.3 Conductors: copper, gold, silver; insulators: rubber, plastic, dirt

12.4 If you have excess charge on you, when you touch a piece of equipment the charge will flow through the device (as a current), and most electronics are rated at very low currents.

12.5 6.25×10^{18} protons

12.6 Enough energy to remove one or more electrons

12.7 See figure C-12.7.

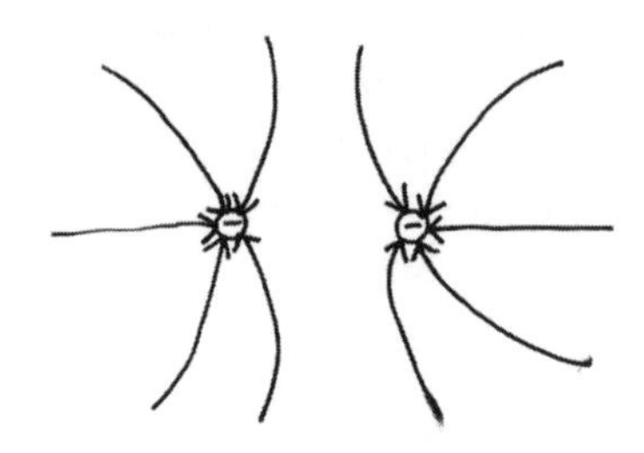

Figure C-12.7

12.8 See figure C-12.8.

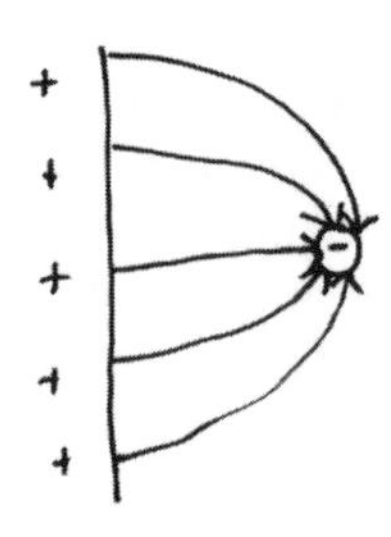

Figure C-12.8

12.9 1.5 V = change in EPE/1 C; change in EPE = 1.5 J

12.10 PD = 10 J/0.5 C = 20 V

Chapter 13

13.1 An alternating current flows in two directions. A direct current flows in one direction. Home outlets carry AC power.

13.2 The chemical potential energy stored in a battery is finite, and once the battery has done a certain amount of work, it must be recharged or disposed of.

13.3 No; a piece of wire can be neutral, but the negative charges are just in motion

13.4 (a) $V = IR$; $1.5\text{ V} = I(150\ \Omega)$; $I = 0.01$ A, or 10 mA; (b) $V =$ IR; $9\text{ V} = (0.1\text{ A})R$; $R = 90\ \Omega$

13.5 $V = IR$; $120\text{ V} = I(100\ \Omega)$; $I = 1.2$ A

13.6 Insulators; no; they conduct heat and electrons well for the same reasons.

13.7 $75\ \Omega$; $V = IR$; $V = 0.01\text{ A}(75\ \Omega) = 0.75$ V

13.8 See figure C-13.8; $150\ \Omega$; $9\text{ V} = I(150\ \Omega)$; $I = 60$ mA

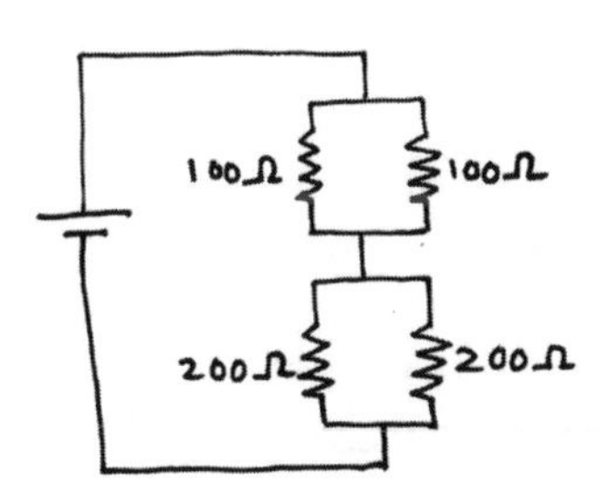

Figure C-13.8

13.9 If the bulbs are connected in parallel (as described in the text), then slightly more current should flow through each bulb, so they should burn more brightly.

13.10 Each device draws more current, and there is a limit to the amount of current a circuit can take.

Chapter 14

14.1 Yes; the magnetism in the steel of the door is induced, and thus will be attracted to either side.

14.2 See figure C-14.2.

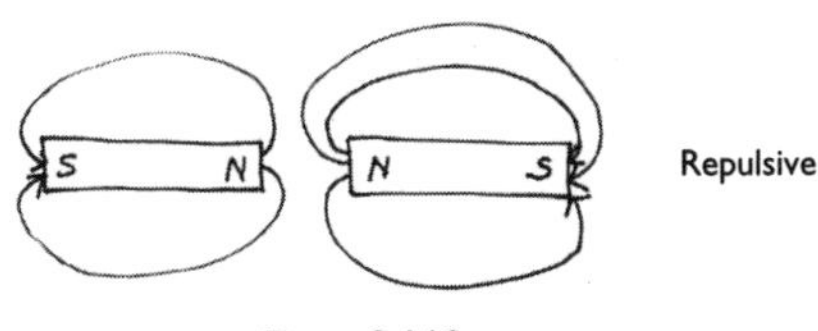

Figure C-14.2

14.3 See figure C-14.3.

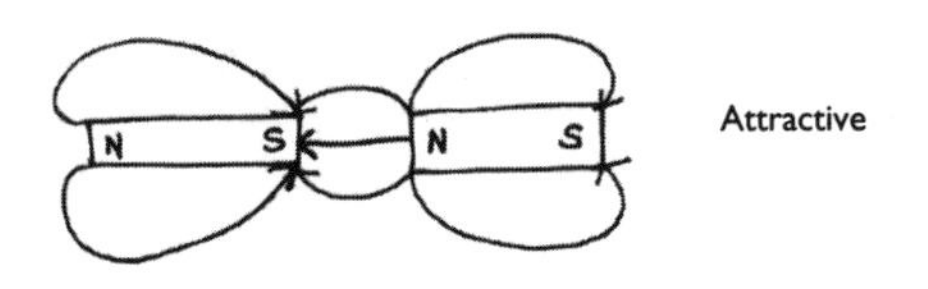

Figure C-14.3

14.4 Hydroelectric dams require large rivers, and there are only a small number of rivers large enough to generate significant amounts of power. Coal is plentiful, and although coal-burning power plants are a major source of pollution, they are relatively cheap and safe to operate.

14.5 $F = (1.60 \times 10^{-19}\text{ C})(0.5)(3 \times 10^{8}\text{ m/s})(1\text{ T}) = 2.4 \times 10^{-11}$ N (into the page)

14.6 Yes. As the magnet moves, the strength of the field in the vicinity of the wire changes, inducing a current in the loop. This induced current will light the bulb.

14.7 $I_p = 1\text{ A}(25{,}000/50) = 500$ A

14.8 A changing magnetic field induces a current. If the induced current were to generate a field of the same type, then a larger current in the same direction would be generated, and so on, leading to a runaway situation.

Chapter 15

15.1 There is evidence all around us; one of the best pieces of evidence is spectral lines from elements. The lines are produced at characteristic wavelengths.

15.2 $E = hc/\lambda$; $E = (6.63 \times 10^{-34}\text{ J s})(3 \times 10^{8}\text{ m/s})/0.02\text{ m} = 9.9 \times 10^{-24}$ J

15.3 $E = hc/\lambda$; $E = (6.63 \times 10^{-34}\text{ J s})(3 \times 10^{8}\text{ m/s})/(10 \times 10^{-6}\text{ m}) = 2.0 \times 10^{-20}$ J, or 10,000 times more energy than a 2-cm photon

15.4 Because an accelerating electron (moving in a curved path) would emit energy continuously and would collapse into the positively charged nucleus

15.5 $n = 1$; there is no highest principal quantum number

15.6 It means enough energy has been added to an atom or to a group of atoms to remove one or more electrons from the valence shell.

15.7 Use $E = hc/\lambda$; $(10.1\text{ eV})(1.6 \times 10^{-19}\text{ J/eV}) = (6.63 \times 10^{-34}\text{ J s})(3 \times 10^{8}\text{ m/s})/\lambda$; thus, $\lambda =$ 120 nm (this is an ultraviolet photon)

15.8 It involves probabilities and not absolute quantities. It contains an admission about what we cannot know.

Chapter 16

16.1 Because it established experimentally that the speed of light is absolute

16.2 The Lorentz factor is $1/\sqrt{1-(0.98)^2} = 5.0$

16.3 $m/m_0 = 1/\sqrt{1-(0.95)^2} = 3.2$ times

16.4 $L = L_0$/Lorentz factor = 4.5 LY/7.1 = 0.63 LY

16.5 Lorentz factor = 7.1; Earthbound observer sees the distance as 4.2 LY and the speed as 0.99c; therefore, time $t = d/v = 4.2(c/0.99\ c)$ = 4.24 years; the total apparent time to the Earthbound observer is thus 4.24 + 2 + 4.24 years = 10.5 years. The traveler sees only 4.24/7.1 = 0.60 years to have passed while traveling there, 2 years at the destination, and 0.60 years to return, for a total of 3.2 years.

16.6 $E = mc^2$; $E = (75\ \text{kg})(3 \times 10^8\ \text{m/s})^2 = 6.8 \times 10^{18}$ J; and you thought you had no energy!

16.7 $E = mc^2$; $E = (1.67 \times 10^{-27}\ \text{kg})(3 \times 10^8\ \text{m/s})^2 = 1.5 \times 10^{-10}$ J

16.8 General relativity is general in the sense that it applies to accelerating reference frames; special relativity treats the special case of relative motion at constant velocity.

Chapter 17

17.1 Fission is the spontaneous splitting of heavy nuclei to form lighter ones; fusion is the combination of lighter nuclei to make heavier ones. Fusion powers the Sun.

17.2 An alpha particle contains 2 protons and 2 neutrons; it has a charge of +2. Helium-3 contains 2 protons and 1 neutron. Helium-4 contains (like an alpha particle) 2 protons and 2 neutrons; Helium-4 is a more abundant isotope of elium.

17.3 If there were no strong nuclear force, the Coulomb repulsion of positively charged protons would tear the nucleus apart; atoms heavier than hydrogen could not exist.

17.4 In a large nucleus containing many protons, it is possible that protons might get far enough away from one another (on opposite sides of the nucleus) that the attractive strong force is weaker than the repulsive Coulomb force.

17.5 125 kg will be left after 40,000 years (4 half-lives).

17.6 Half-life of carbon-14 is 5730 years. 1/32 is $(\frac{1}{2})^5$, so that 5 half-lives have passed, or 5 × 5730 = 28,650 years.

17.7 The gravitational force

17.8 $E = mc^2$; $E = (2 \times \text{mass of electron})c^2 = (2)(9.11 \times 10^{-31}\ \text{kg})(3 \times 10^8\ \text{m/s})^2 = 1.6 \times 10^{-13}$ J

17.9 $E = mc^2$; $E = (2 \times \text{mass of proton})C^2 = (2)(2000)(9.11 \times 10^{-31}\ \text{kg})(3 \times 10^8\ \text{m/s})^2 = 3.3 \times 10^{-10}$ J, or about 2000 times more energy

GLOSSARY

absolute zero
Temperature at which gases cease to exert pressure; 0 K, or –273°C

absorption lines
Dark lines superimposed on a continuous spectrum corresponding to electronic transitions in excited atoms; each element has characteristic absorption lines

acceleration
Rate of change of velocity, or the change in velocity divided by the time it takes for the change to occur

acceleration due to gravity
Constant rate of acceleration of falling bodies, ignoring air resistance or friction; 32 ft/s, or 9.80 m/s

alloy
Solid mixture containing two or more metals, or a metal and other elements

alpha particles
Positively charged particles consisting of two neutrons and two protons

alternating current (AC)
Type of current that naturally results from the continued turning of a coil (e.g., an electromagnet) in a fixed magnetic field; the type of current in your house

ammeter
Instrument for measuring the flow of electricity

ampere (amp)
Rate of flow equal to one coulomb of electric charge per second

amplitude
"Height" of a wave, measured as a displacement from a zero level

angular acceleration
Change in angular velocity with time of a rotating object

angular momentum
Product of an object's moment of inertia and its angular velocity; rotational equivalent of linear momentum

angular motion
Rotational displacement (in radians or degrees) of a rotating object

angular velocity
Change in angular position with time of a rotating object

area
Measurement of the two-dimensional size of an object; expressed in units of length squared (e.g., ft^2, m^2)

atom
Once thought to be the smallest component of matter; now thought to be simply the smallest unit of an element

atomic mass number
Total number of neutrons and protons in the nucleus of an atom

atomic number
Number of protons in the nucleus of an atom

barometer
Instrument used to measure air pressure

beta particles
Another name for electrons

blueshift
Change in the frequency of light or sound waves and other wave phenomena that results when an observer and a source are moving toward each other

boiling
Formation of vapor bubbles throughout a liquid being heated at the point when the pressure of the vapor equals atmospheric pressure

British thermal unit (Btu)
Quantity of heat (thermal energy) needed to raise the temperature of one pound of water one degree Fahrenheit

buoyancy
Tendency of an object to rise (float) when submerged in a liquid

calorie
Quantity of heat required to raise the temperature of one gram of water one degree Celsius; basic unit of thermal energy in the metric system; kilocalorie (kcal) is used in dealing with large quantities of heat

capacitance
Rating, stated in ohms, of the ability of a nonconductor (dielectric) to store charge when there is a difference in potential between its opposite surfaces

Celsius scale
System of temperature measurement in which the ice point is designated as 0° and the steam point as 100°; formerly called *centigrade* system

center of gravity
The position in an object where its mass can be imagined to be concentrated; the balance point of an object

centimeter
Metric length measurement equal to 0.01 m

centrifugal force
Force that acts outwardly (away from the center) in a rotating object

centripetal force
Force that acts inwardly (toward the center), tending to keep an object moving in a circular path

charge
Property of matter that is a measure of its excess or deficit of electrons

chemical potential energy
Energy stored in chemical bonds

coefficient of linear expansion
Measure of the incremental expansion or contraction in length of an object with temperature

commutator
Split conductive cylinder that is used to reverse the direction of current in a coil of wire located in a magnetic field

compound
Substance that is the combination of two or more elements

concave lens
Lens that causes incoming parallel rays of light to diverge

concave mirror
Mirror that has an inward curvature facing incoming light rays (e.g., the inside of a spoon)

conduction
Transfer of molecular motion through a substance by collisions; one way in which heat passes from one material to another

conductor
Material that allows electric charge to pass freely, with little resistance

convection
Mass movement of a heated liquid or gas

convex lens
Lens that causes incoming parallel rays of light to converge

convex mirror
Mirror that has an outward curvature facing incoming light rays (e.g., the outside of a spoon)

coulomb
SI unit of electric charge; equal to the quantity of electricity moved by a current of 1 A in 1 s

critical angle
Incident angle—measured from the normal—that if exceeded will cause rays of light to experience total internal reflection at a boundary

deceleration
Negative acceleration; (see *acceleration*)

decibel (dB)
Unit for measuring the intensity of sound; the decibel scale is logarithmic

degree
Unit for measuring temperature on various scales

density
Mass of a substance divided by its volume

deuterium
Isotope of hydrogen consisting of one proton and one neutron

diffraction
Bending of waves as they pass near the edge of an obstacle or through a small opening

diffraction grating
Series of many small, equally spaced, parallel slits (or lines) inscribed on a polished surface, used to produce spectra

diffuse reflection
Reflection from a rough or irregular surface

direct current (DC)
Type of current in which the charge flows continually in one direction; the type of current in handheld electronics

dispersion
Separation of light into its component colors; the separation of radiation into its components by wavelength or some other characteristic

displacement
The weight or volume of a fluid that has been replaced by a submerged object

Doppler effect
Change in the frequency of light or sound waves, and other wave phenomena that results when an observer and a source are moving relative to each other; for sound, the increase in pitch as a listener and a source of sound approach each other, and the decrease in pitch as they move away from each other

electric circuit
Complete path of an electric current, including a source of potential difference and usually including various components (e.g., resistors, diodes)

electric current
Charge in motion

electric field
Space in the vicinity of a charged object in which its force or some part of it is exerted

electric force
Attractive or repulsive force between two charged objects; also called the Coulomb force

electric resistance
Opposition of a conductor to the passage of electric current

electricity
Electric current or power; the study of charges in motion

electromagnet
Device for increasing electromagnetic force by means of a soft-iron core

electromagnetic force
Combination of magnetic and electric forces, due to the motion of charged particles

electromagnetic induction
Production of electric current by means of a coil and a magnet

electromagnetic waves
Any waves that result from the motion of charged particles

electromotive force (emf)
A potential difference that causes electric charges to flow

electron
Smallest indivisible particle with negative charge

electrostatic induction
Production of an electric charge in a neutral body by bringing it near a charged one

electrostatics
Study of charges that are not in motion

element
Chemical that cannot be further broken down into component chemicals

emission lines
Bright spectral lines emitted by excited atoms with or without a continuous spectrum as a result of electronic transitions

energy
Ability to do work

equilibrium
Point at which the resultant force acting on an object is zero; rest or balance

evaporation
Escape of molecules from the surface of a liquid

Fahrenheit scale
System of temperature measurement in which the freezing point is 32°F and the boiling point is 212°F

fiber optics
Thin, transparent fibers of glass or plastic surrounded by material with a lower index of refraction that transmit light by total internal reflection; used in many ways, including communication and in medical diagnosis

field lines
Lines indicating the direction of a force field

fission
Splitting of the nucleus of an atom

focal length
Distance from a lens at which incoming parallel rays converge (convex), or from which they appear to diverge (concave)

foot-candle
Unit used to express illumination of a surface that is one foot away from a point source of light of one candle; equal to one lumen per square foot

force
Push or pull exerted on or by an object

force vector
Amount or magnitude of a force plus its direction

free fall
State of an object under the influence of a gravitational force with no counteracting force

frequency
Number of vibrations per second of a wave; the reciprocal of period

fundamental units
Units of measurement that are the basis of our measurement of the universe; the meter (length) and second (time) are examples of fundamental units

fusion
Joining of atomic nuclei to make heavier nuclei

galvanometer
Coil of wire that is free to rotate in a magnetic field; can be used to measure voltage or current

gamma rays
High-energy electromagnetic radiation, shorter wavelengths than x-rays

gas
State of matter that has neither definite volume nor definite shape

generator
Device used to convert mechanical energy into electrical current

gravitational potential energy
Energy associated with position in a gravitational field, or the amount of work an object can perform by returning to its original position

gravity
Attractive force between objects with mass; the curvature of space–time induced by the presence of mass

ground state
Lowest possible energy state of an electron bound to a nucleus

grounding
Loss of charge that occurs when a charged object is connected to a very large body with an almost infinite capacity to provide or absorb electrons

half-life
Amount of time required for half of the mass of a radioactive substance to undergo radioactive decay

heat
Thermal energy that can be transferred between two bodies at different temperatures

heat of fusion
Quantity of thermal energy required to melt a gram of a given substance without producing any change in temperature; characteristic for each material

heat-work equivalent
Rate of exchange between mechanical energy and thermal energy; whenever a given amount of mechanical energy disappears, a fixed quantity of thermal (heat) energy appears in its place

hertz (Hz)
Unit of frequency; 1 Hz equals 1 cycle per second

ideal gas
Simplified model of a gas in which the molecules or atoms can be thought of as individual, point like particles that do not exert intermolecular forces on one another

illumination
Radiant energy falling on any given unit area

index of refraction
Ratio of the speed of light in a vacuum to its speed in a given material; always greater than 1

infrared
Electromagnetic radiation beyond the red end of the visible spectrum, with longer wavelengths than visible light

insulator
Nonconductor; material that impedes passage of electric charge

interference
Summation of two or more waves occupying the same space; can be constructive or destructive

ion
Particle with charge due to the loss or gain of one or more electrons

ionization
Complete removal of an electron from an atom

kelvin
Temperature increment of the Kelvin, or absolute, scale of temperature; formerly called degree Kelvin

Kelvin scale
Temperature scale based on absolute zero (–273°C) Also called *absolute scale.*

kilogram
1000 g; metric mass measurement

kilometer
1000 m; metric length measurement

kinetic energy (KE)
Energy of a moving body

latent heat
Amount of heat required to change the phase of a given mass of a substance

lens
Device, often made of plastic or glass, that changes the direction of rays of light, usually in order to form an image

liquid
State of matter in which the matter has no shape of its own but takes the shape of its container

longitudinal wave
Wave in which the disturbed particles move parallel to the direction in which the wave is advancing

luminous intensity
Brightness of a light source, measured in units of candela (also called *standard candle*)

magnetic field
Region around a magnet in which its effects are exerted

magnetic flux
Total number of lines of force passing through any circuit located in a magnetic field

magnetic induction
Temporary transference of magnetism from a permanent magnet to another material

magnetic poles
The two ends of a magnet, north and south

magnetism
Ability to attract iron and certain other metals with a similar molecular structure

mass
Amount of matter an object or substance contains

matter
In classical physics, anything that takes up space; in modern physics, matter and energy are interchangeable

mechanical energy
Amount of work an object (body) can do

meter
Fundamental unit of length, about 39.37 in.

metric system
System of measurement in which all fundamental units are multiples of 10

molecule
Stable combination of two or more atoms

moment of inertia
Measure of an object's resistance to changes in its rotational state; rotational equivalent of mass

momentum
Result of forces acting on objects with mass; or the product of mass and velocity

neutrino
Uncharged particle with little or no mass; produced in beta decay.

nucleons
Neutrons and protons; components of the atomic nucleus

order of magnitude calculation
Estimate that is approximate but accurate to the nearest power of 10

oscillate
Move back and forth about a center; vibrate

parallel circuit
Connection of electrical components in such a way that current can branch in multiple directions, one through each component in parallel

period
Time required to complete one cycle of a wave; the reciprocal of frequency.

phase
Position in the cycle of a wave

photoelectric effect
Ejection of electrons from a metal surface when light strikes it; a quantum effect

photon
Carrier of the electromagnetic force

physics
Study of matter, energy, and the laws governing their interactions

positron
Antiparticle of the electron, with the same mass and opposite charge

potential difference
Difference in electric charge between two objects; a charge will tend to move from the area of higher potential to the area of lower potential

power
Rate of doing work

pressure
Amount of force exerted by an object on the area of the surface on which it acts

primary colors
The three colors from which every other color can be formed: red, green, and blue-violet

quanta
Discrete packets in which many forms of energy travel

quantum mechanics
Study of quanta

quark
Building block of neutrons and protons

radiation
Movement of wave forces out from a center; also, the process in which heat is transferred by waves traveling through space

radioactive decay
Spontaneous disintegration of the nuclei of sufficiently heavy atoms

real image
Image that can be brought to focus

redshift
Change in the frequency of light or sound waves, and other wave phenomena, that results when an observer and a source are moving away from each other

reflection
Change in the direction of rays that bounce off a boundary between two substances (e.g., air and glass)

refraction
Change in the direction of rays passing through a boundary at an angle to the normal

resistance
Capacity of an object or material to impede motion; also, the capacity of a material to impede the motion of charge

resonance
Process by which sound vibrations build up

rest frame
Frame of reference in which an object is not in motion

resultant
Single vector of definite magnitude and direction that is the combination of a number of other vectors; multiple forces are often thought to add to a single resultant vector

scalar
Measured quantity that has size but no direction; mass is a scalar quantity

scientific notation
Shorthand way to refer to very large and very small numbers; numbers in scientific notation have a coefficient, a base, and an exponent

second
Basic unit of time in both the English and the metric systems

semiconductor
Solid-state device having medium resistivity used to transmit and amplify electronic signals

series circuit
Connection of electrical components in such a way that the same current flows through each component

solenoid
Coil of wire that can carry current; used in transformers

solid
State of matter that has a definite shape and volume

sonic boom
Sound produced by the shock wave generated by an object moving faster than the speed of sound as it compresses the air ahead of it

special relativity
Model of the universe that takes into account the finite speed of light in all reference frames

specific heat
Amount of heat required to raise the temperature of 1 g of a substance 1°C

spectroscope
Device used to view a spectrum

spectrum
Sequence of colors (wavelengths) that results when light is dispersed through a prism

speed
Rate at which something moves

standard candle
Unit of measurement of the luminous intensity or brightness of a light source (also called *candela)*

standing wave
Wave that has fixed points (nodes) and moving points (antinodes); wave that appears "frozen" in a medium, like a string

static electricity
Electric charge resting on an object

strong nuclear force
Attractive force between nucleons

superconductor
Substance that at low temperature has almost no resistance to the passage of current

temperature
Degree of hotness or coldness of an object or an environment; a measure of the average velocity of particles in a substance

thermal energy
Heat energy

thermodynamics
Study of the relationship between heat and mechanical energy

time
Continuum along which events move from the past through the present and into the future; not an absolute according to special relativity

time dilation
Relativistic effect in which an observer at rest with respect to another moving at constant velocity will measure that more time has passed than the moving observer will

torque
Ability of a force to produce rotation; rotational analog to force

total internal reflection
Reflection rather than refraction of all incident rays striking the boundary between two substances

transformer
Device that changes voltage and current values in AC circuits

transverse wave
Wave in which the disturbed particles move perpendicular to the direction in which the wave is advancing

ultrasonic frequencies
Frequencies above the hearing range of humans, especially those of several hundred thousand hertz

ultraviolet
Electromagnetic radiation falling just beyond the violet end of the visible spectrum; shorter wavelengths than the violet

vacuum
Space devoid of matter

vector
Measured quantity that has both a magnitude and a direction; weight, for example, is a vector quantity

velocity
Speed measured in a particular direction (a vector quantity)

virtual image
Image that cannot be brought to focus

volt
Measure of potential difference equal to 1 J/C

voltmeter
Instrument for measuring voltage that passes current through a rectangular coil having a high resistance

volume
Amount of three-dimensional space occupied by matter

watt
Measurement of power; working rate of 1 J/s

wavelength
Separation between successive crests of a wave

weak nuclear force
Force responsible for the spontaneous decay of neutrons

weight
The pull of Earth's gravity on an object

work
Transfer of energy to an object by the application of a force over some distance

x-ray
Waves of shorter length than ultraviolet, often emitted by radioactive substances

INDEX

NOTES